CLIMATE
HOPE

CLIMATE
HOPE

On the
FRONT LINES
of the FIGHT
AGAINST COAL

TED NACE

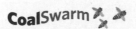

CLIMATE HOPE: On the Front Lines of the Fight Against Coal
Ted Nace

Published by
CoalSwarm
1254 Utah Street, San Francisco, CA 94110
415-206-0906 info@cmNOW.org

Adapted Material
Chapter 3, "Inside the Swarm" is adapted from Ted Nace, "Stopping Coal In Its Tracks," *Orion Magazine*, January 2008, copyright © 2008 by The Orion Society; Chapter 9, "The Education of Warren Buffet," is adapted from Ted Nace, "The Education of Warren Buffett, *Gristmill*, April 15, 2008, copyright © 2008 by Grist Magazine; Appendix A, "Protests Against Coal," is adapted from "Nonviolent Direct Actions Against Coal," "Nonviolent Direct Actions Against Coal: 2004–2007," "Nonviolent Direct Actions Against Coal: 2008," and "Nonviolent Direct Actions Against Coal: 2009," all published on the SourceWatch wiki. Appendix B, "Coal Plants Cancelled, Abandoned, or Put on Hold," is adapted from "Coal Plants Cancelled, Abandoned, or Put on Hold: 2007," Coal Plants Cancelled, Abandoned, or Put on Hold: 2008," and "Coal Plants Cancelled, Abandoned, or Put on Hold," all published on the SourceWatch wiki. SourceWatch content is available under GNU Free Documentation License 1.3, a form of copyleft that encourages sharing. For information see http://SourceWatch.org.

Cover design: Mimi Heft
Cover photo: Kurt Mann
Text design: Mimi Heft
Page layout: Mimi Heft and David Van Ness
Copyediting: David Sweet
Errors & Omissions: Ted Nace

ISBN-13: 978-0615314389
ISBN-10: 0615314384

1 2 3 4 5 6 7 8 9

Printed in the United States of America

In memory
Lovina Anne Kelley Nace
1925–2009

_____

Contents

Introduction

■

ON THIS JUNE DAY in 2009 the West Virginia mountains watched quietly as mortals squabbled over their fate. Along the scorching hot asphalt of state highway 3 next to the Massey coal processing plant, flanked by officers holding nightsticks, anti–mountaintop removal protesters walked toward the plant gate to deliver a letter to coal baron Don Blankenship. In the south lane a German shepherd police dog, its leash gripped by an officer, snarled menacingly while its handler shouted at passers-by to step back. Across the road, a crowd of sweating, red-faced coal miners shouted threats and obscenities and chanted "Massey! Massey!" Suddenly a shrieking woman emerged, striking one of the demonstrators in the face before being seized by police.

In the eye of the turbulence the world's most prominent climate scientist, Dr. James Hansen, stood calmly as police officers cinched plastic handcuffs onto his wrists and placed him in a patrol car. For a time the car remained motionless, held in check by the packed throng; then it slowly pushed its way up the road toward the county detention center.

Earlier in the day, Dr. Hansen had delivered a speech in which he stated that climate change, left unchecked, would drive half or more of all the species of plants and animals on

Planet Earth into extinction. The rally had been invaded by dozens of miners who had been given the day off to heckle the speakers. Arriving on loud motorcycles, many with body-builder physiques and shaved heads, the miners had formed an intimidating wall directly behind the speakers' platform. They shouted insults, unplugged the power cord to the sound equipment, and set off air horns directly into the ears of blue-grass musicians entertaining the rally. When 94-year-old Ken Hechler, a former congressman and West Virginia secretary of state, told the rally about his decades-long efforts to halt mountaintop removal, the miners shouted at him to get back into his wheelchair.

As I watched these scenes of chaos, it was obvious what motivated both sides of the controversy. On one side were West Virginians whose families had long treasured these beautiful mountains, in some cases for over two hundred years. Most Americans, faced with the destruction of their homes, would fight just as hard. On the other side were workers who feared for their livelihoods and their families. Though they had been manipulated into serving as thugs for an unscrupulous corporate boss, their personal concerns were no less valid.

But the head of climate research for NASA? Here was an insider with consummate access to the halls of federal power. Why should such an individual find it necessary to come to an obscure town in West Virginia and face an angry mob, just to deliver a letter? Was the message really so urgent? What it is about coal—just one among many sources of pollution—that would motivate a climate scientist to go to jail? Why had a movement coalesced around these issues, and what had that movement accomplished so far?

These are the questions that this book attempts to explore.

■

ONE

The 80% Solution

■

DANA MILBANK SOUNDED MYSTIFIED, or at least surprised. It was a typically lovely summer day in 2008, and the longtime political reporter for the *Washington Post* had been following NASA's chief climate scientist, James Hansen, as he made the rounds of media and government in Washington, D.C.: Diane Rehm's talk show, congressional committees, and the National Press Club. At age sixty-seven, Hansen, who works in New York City as the director of NASA's Goddard Institute for Space Studies, was a familiar face in the capital. He had been coming to the Hill for at least twenty-five years to talk with bureaucrats, reporters, activists, legislators—indeed, with anyone who would hear him out—and in all that time the core message, while disturbing in its implications, had been remarkably consistent. But today, it seemed to Milbank, the message had developed a distinct new wrinkle. At least the reporter couldn't recall hearing it before.

With his balding head and slow-paced Iowa deadpan, Hansen could have stood in for practically any role in a Thornton Wilder play set in small-town America—repairman, farmer,

high school basketball coach—but the persona was decep-
tive. For one thing, it masked a world-class intellect. Having
begun his career studying the atmosphere of Venus and other
planets, Hansen had gone on to apply that expertise to the
study of Earth's own atmosphere and had become one of the
early pioneers of the ultrasophisticated "general circulation
models" that are now the gold standard for supercomputer
forecasting of climate change in the United States, Europe,
Japan, and elsewhere. His work has won him acclaim within
the scientific community, including membership in the elite
American Academy of Sciences and award of the Carl-Gustaf
Rossby Research Medal, the highest honor bestowed by the
American Meteorological Society.

Of course, one would expect NASA's top climatologist to
have that sort of résumé. The lack of conscious polish added to
Hansen's credibility—and even a charisma of sorts—that had
served the scientist well over the years. But even though people
were able to see the brilliance behind the self-effacement, what
they didn't tend to see was more in the realm of temperament:
a distinct proclivity for inductive leaps that made some of Hansen's more cautious colleagues uncomfortable at times. Climate
science has always had to struggle in the popular media with
the question, If you can't even predict tomorrow's weather, what
makes you think you can say anything meaningful whatsoever
about conditions at the end of the century? That objection
overlooks a simple difference between meteorology and climatology. While both make predictions about Earth's complex
atmospheric system, climatology's job is actually easier because
it deals not with particular weather events but with long-term
trends in temperature, precipitation, and other measurable
features. Thus, while neither a predicted temperature rise over

a hundred-year period nor the appearance of a tornado in the next twenty-four-hour period can be predicted with 100 percent accuracy, a forecast about the former can be made with a higher degree of certainty than a forecast about the latter.

Yet even though climate shows greater regularity than weather, climate scientists remain a wary bunch, and Hansen's penchant for bold hypothesizing was unusual. In 1981 he published a paper in *Science* predicting that the 1980s would be an unusually warm decade worldwide and that the 1990s would be even warmer. Both predictions turned out to be correct. In 1988 he told Congress that by the end of the century, unambiguous signals of a warming trend would emerge worldwide out of the general noise of temperature data. In fact, the twelve-year period from 1997 to 2008 included the ten hottest years on record. In 1990 Hansen bet climatologist Hugh Ellsaesser $100 that one of the following three years would be the hottest on record. Hansen agreed to a tough definition: to be considered "hottest," the year would have to hit new highs on three different planetary measures: land surface temperature, ocean surface temperature, and temperature of the lower atmosphere. Within six months, all three measures broke records, and Ellsaesser had conceded the bet.

None of Hansen's predictions was reckless: all were supported by the models. Yet had the rising temperatures failed to appear, his credibility would have been severely damaged. Among those who had developed a deep respect for Hansen's intuition was physicist Mark Bowen, who believed that the scientist's willingness to go out on a limb derived directly from his commitment to the scientific method. Hansen had written, "The way I look at it, the great fun in science is that you get to reason about how things work, leading you to make predictions that test your understanding. The predictions that you make had better include some that are wrong or you are not pushing the envelope of scientific understanding."

In other words, it wasn't so much that Hansen had no concern about ever being wrong—every scientist wants to be right—but that he was driven more by an intense desire to break new ground. This made him surprisingly unperturbed by "climate skeptics," whom Hansen often credited with having helped him strengthen his theories.* ABC News reporter Bill Blakemore commented, "He's transparently full of integrity.... You get the feeling that this is a guy to whom it wouldn't even occur to lie."

"The work that he did in the seventies, eighties, and nineties was absolutely groundbreaking," physicist and historian Spencer Werrt told *New Yorker* reporter Elizabeth Kolbert. He added, "It does help to be right."

A quarter century after Hansen's original predictions, the reality of global warming and the role of human activity in that warming had become well established within the scientific community. Princeton climatologist Michael Oppenheimer told Kolbert, "I have a whole folder in my drawer labeled 'Canonical Papers.' About half of them are Jim's."

In 2007 the national science academies of Brazil, Canada, China, France, Germany, Italy, India, Japan, Mexico, Russia, South Africa, the United Kingdom, and the United States

---

* Hansen himself might actually be the original global warming skeptic, having written his doctoral thesis on a hypothesis about the cause of Venus's high temperatures that countered the dominant theory of the time. Whereas most scientists (including Carl Sagan) believed that greenhouse gases such as carbon dioxide were responsible for the phenomenon, Hansen attempted to prove that a blanket of fine dust was responsible. Once satellite probes demonstrated that high concentrations of greenhouse gases were indeed present on Venus, Hansen willingly ceded the point.

jointly endorsed the main conclusions of the body of research that Hansen, more than any other single scientist, had been responsible for developing. The thirteen national academies issued the following joint statement: "It is unequivocal that the climate is changing, and it is very likely that this is predominantly caused by the increasing human interference with the atmosphere. These changes will transform the environmental conditions on Earth unless counter-measures are taken."

Having been vindicated by the course of events, Hansen was not content to rest on his laurels. In accordance with NASA's stated mission "to understand and protect our home planet," he had made it his brief to inform policy makers and the public about the consequences of global warming in terms of sea level rise, extreme weather events, drought, and species extinction, in effect conducting a long-running seminar that he continually updated and extended. Going even further, he had led a coordinated research effort by an international assembly of climate scientists to determine the threshold concentration of greenhouse gases that would constitute "dangerous anthropogenic interference with the climate system" and to identify energy strategies that would most effectively avert such dangerous concentrations.

In discussing the results of this new research, Hansen was now back in the public arena, with a message that Dana Milbank, in his column the next day recounting his time with Hansen, described as "rather counterintuitive." The message wasn't so much about the contents of the atmosphere or the oceans—the two systems normally studied by climate scientists—but of the Earth's crust, specifically the fossil fuels contained in that crust.

Up until now, most of the efforts expended by environmentalists and others to limit fossil energy use had focused on limiting overall emissions, regardless of the source. That made sense. After all, a molecule of carbon dioxide does not know whether it came from the burning of oil, natural gas, or coal. There's no difference between "oil carbon," "natural gas carbon," and "coal carbon."

But now Hansen was maintaining exactly the opposite: not all carbon is created equal; there is a difference. Specifically, Hansen was saying the carbon from oil and gas, despite its ubiquity and despite all the many efforts to limit its emissions—from attempting to legislate fuel efficiency standards to moving consumers away from gas-guzzling SUVs and toward Priuses and other fuel-efficient vehicles—was less important than the carbon from coal, which is mainly used as fuel in power plants.

Milbank summed up the new twist in Hansen's message this way: "[T]the biggest worry isn't what we put in our cars, it's what we put in our power plants." Or, to boil the message down even further:

Want to stop global warming?
Forget oil and gas.
Stop coal.

Forget oil and gas? To anyone who had been following the course of energy policy in the United States over the past several decades, beginning with the energy crisis of 1973, which had been triggered by the boycott of the United States organized by the Organization of Petroleum Exporting Countries (OPEC), oil and gas had always been the center of the story. Coal had virtually been ignored.

In fact, to most Americans, "energy crisis" and "oil crisis" had always been nearly synonymous. Now "climate crisis" was the focus of concern, and once again oil was the focus of most people's thinking. Not surprisingly, SUVs, Hummers, and other gas guzzlers had become the poster villains for environmentalists. For anyone who wanted to demonstrate concern about global warming, driving a Prius had become a badge of merit.

As for coal, most people were barely aware of its role in the mix of energy sources. The word evoked images from an earlier time: railroad steam engines, coal cellars in hundred-year-old houses. Few Americans had ever seen a coal-fired power plant; fewer still had laid eyes on a coal mine. If anything, coal was touted as a savior. With an estimated 29% of world reserves, America was known as the "Saudi Arabia of coal."

Now Hansen was saying that the abundance of coal, far from being a cause for comfort, was actually our worst problem. In fact, he was willing to put a number on its importance. Ending emissions from coal, he said, "is 80% of the solution to the global warming crisis."

Hansen's reasons for emphasizing coal were fourfold:

■ **First,** as shown in figure 1, the amount of carbon remaining in the ground in oil and gas reserves is much smaller than the amount of carbon contained in coal reserves.

■ **Second,** coal is the most carbon intense of the fossil fuels. Producing a kilowatt-hour of electricity from coal produces about 2.4 pounds of carbon dioxide, while producing a kilowatt-hour of electricity from natural gas produces about 1 pound of carbon dioxide. While coal produces half of the electricity used in the United States, it is responsible for 80 percent of the carbon dioxide released by electric utilities.

■ **Third,** coal consumption is far more concentrated than the use of other fossil fuels. A mere six hundred large coal-burning power plants account for nearly all coal usage,

FIGURE 1 HISTORICAL FOSSIL FUEL EMISSIONS AND REMAINING RESERVES

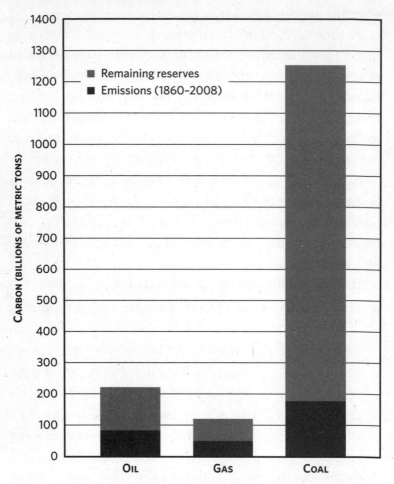

Source: Adapted from James Hansen et al., "Target Atmospheric CO_2: Where Should Humanity Aim?" *Open Atmospheric Science Journal* (2008): page 11. Estimates for remaining oil, gas, and coal reserves are from Intergovernmental Panel on Climate Change, *Climate Change 2001: Mitigation*, B. Metz et al., eds. (New York: Cambridge University Press, 2001).

in contrast to the tens of millions of cars, trucks, planes, homes, businesses, and factories that burn oil and gas. Thus, reducing emissions from coal is a far simpler task.

■ **Fourth,** production of oil and gas is primarily located in countries that American domestic energy policy has little or no ability to control. Any reduced consumption by the United States might well be offset by increased consumption in other countries. In contrast, our ability to control the consumption of coal is substantial, since the United States leads the rest of the world in the size of its coal reserves.

It would have been easy to dismiss Hansen if he were the only scientist making such assertions. But numerous other climate scientists backed up the conclusions he was now talking about. For example, he had coauthored his most recent paper, "Target Atmospheric CO_2: Where Should Humanity Aim?" with nine other prominent researchers: Makiko Sato and Pushker Kharecha of NASA and Columbia University, David Beerling of the University of Sheffield, Robert Berner and Mark Paganini of Yale University, Valerie Masson-Delmotte of the University of Versailles, Maureen Raymo of Boston University, Dana L. Royer of Wesleyan University, and James C. Zachos of the University of California. All had joined Hansen in endorsing the paper's central conclusion:

> Present policies, with continued construction of coal-fired power plants without CO_2 capture, suggest that decision-makers do not appreciate the gravity of the situation. We must begin to move now toward the era beyond fossil fuels. Continued growth of greenhouse gas emissions, for just another decade, practically eliminates the possibility of near-term return of atmospheric composition beneath the tipping level for catastrophic effects. The most difficult task, phase-out over the next 20–25 years of coal use that does not capture CO_2, is herculean, yet feasible when compared with the efforts that went into World War II.

The stakes, for all life on the planet, surpass those of any previous crisis. The greatest danger is continued ignorance and denial, which could make tragic consequences unavoidable.

The intensity of the warnings by Hansen and his fellow climate scientists made me recall a brief conversation with a friend, some months earlier. We were in the kitchen of a rental cabin overlooking the Pacific Ocean, having just arrived with two hatchbacks filled to the brim. We were unpacking our bags of bread and vegetables, bottles of wine, chocolate bars and cookies, beach toys, towels, children's storybooks, and all the other things one brings along on a weekend for two families trying to escape the hectic demands of everyday life.

I knew that Michael was an expert on energy and climate and that he worked for the Stockholm Environment Institute, a think tank that advises governments and nonprofit organizations around the world. Beyond that, I had only the vaguest notion of what sort of work he actually did. This was to be a weekend for us to go off with our families and have a little time away from the daily grind, so I didn't expect topics like global warming to be major topic of conversation. But I was curious to know what he had been up to, so I thought I'd ask him a routine question. "How's your climate work going?" I asked.

As he looked at me, I could see him thinking whether to give me the ten-second answer or the thirty-minute answer.

He said, "Most scary things you hear about are probably hype. If you talk to a real expert, you find that journalists have taken something and blown it out of proportion. But with climate it's the opposite. The closer you get to the experts, the more panic you see."

This wasn't the answer I had expected. I thought he'd tell me something more along the lines of "Well, it's a serious problem, but we're working on it and there've been a lot of breakthroughs." I had thought the climate problem was something very gradual, slow—a remote danger.

I mumbled something like "Hmm, that sounds bad" and then quickly changed the subject to something more pleasant. Despite the alarming message, Michael's words left me unaffected. What could I do about such a massive situation? The scale was simply too big, too overwhelming.

That sense of doomed inevitability stayed with me until I encountered Hansen's prescription for solving climate change. What made Hansen's message about coal compelling was that it not only named the primary *driver* of climate change but also identified a potential *solution.* A paper coauthored by Hansen and fellow climatologist Pushker Kharecha* explored the question of what would happen if coal use was phased out but efforts to rein in oil and gas usage proved unsuccessful. Could we win the climate war if we *just* won the coal war?

Though cautiously framed, the conclusion of Hansen and Kharecha was: *Yes.* Due to the more limited size of remaining oil and gas reserves, the two scientists concluded that atmospheric carbon dioxide levels could potentially peak at somewhere between 422 and 446 parts per million before gradually declining, a scenario that would not prevent all global warming—that possibility is no longer an option—but that might well head off more dangerous outcomes.[†]

The idea that the climate change could be addressed by something as straightforward as phasing out coal intrigued

---

[*] P.A. Kharecha and J.E. Hansen, "Implications of 'Peak Oil' for Atmospheric CO_2 and Climate," *Global Biogeochem. Cycles,* 22, 2008.

[†] An important caveat was that use of unconventional fossil fuel sources such as tar sands and oil shale would also have to be avoided.

me, since that did not strike me as an impossible goal. In contrast to the self-defeating notion that climate change can not be stopped by any means short of a wholesale dismantling of industrial civilization, Hansen's message could only be described as hopeful and optimistic.

This is doable, I thought. I knew that there are ways to generate electricity that don't involve burning coal. I also knew that besides the paramount danger of climate change, coal has many other problems, from the ravages of strip mining to the health effects of sulfur dioxide and heavy metal emissions. I decided it was time to learn more.

■

T W O

151 Time Bombs

■

IN THE SPRING OF 2007, a bureaucrat at the U.S. Department of Energy named Erik Shuster put the finishing touches on a routine document, then posted it on the Internet. The document listed 151 coal plants in various stages of completion, from initial proposal to operation.

Shuster had no idea that this number—151—would turn out to be the flash point for a grassroots movement to stop the plants. This latest tally of coal plant projects in the works was simply a routine update to a tracking report that his department had been quietly posting for over five years. I stumbled on Shuster's document while surfing the Web, looking for more information on the wave of proposed new coal-fired power plants. It was one of many stray pieces of information I came across about coal, saving it in my Web browser as I Googled here and there.

In addition to such general pieces of information, I was interested in finding out what the big environmental groups were doing in support of James Hansen's call for a moratorium on new coal plants. I jumped from Web site to Web site—National

Wildlife Federation, Natural Resources Defense Council, Environmental Defense Fund—but except for the Sierra Club, not one of these large national groups was rallying its members in support of Dr. Hansen's call for a nationwide freeze on new coal plants.

This lack of action perplexed me. Here was America's top climate scientist—backed up by nine colleagues—announcing, in effect, a five-alarm fire and laying out in detail where to direct the hoses. Yet it struck me that the environmental establishment as a whole was responding like a fire crew on a coffee break.

A case in point was the Environmental Defense Fund (EDF). With half a million members, hundreds of staff, offices in eleven cities, and revenues of $89 million a year EDF is one of the powerhouses of the environmental movement. Yet EDF's Web site said not a word about the need for an immediate moratorium on new coal plants. Indeed, on EDF's blog the group's chief scientist, Bill Chameides, was claiming that "there are clean coal technologies that will allow us to use our huge coal reserves without harming the climate." (I was soon to learn that such optimism about "clean coal" was not shared by most grassroots activists.)

Another group, the National Wildlife Federation, seemed to recognize the immense threat posed by global warming. The group's annual report expressed the matter clearly: "National Wildlife Federation is dedicated to confronting global warming as the most urgent threat to our mission of protecting wildlife for our children's future." Yet in that same report, the word *coal* did not appear even once.

Checking the Web sites of other major groups, including Nature Conservancy, Wilderness Society, and Audubon Society, I found much the same story. None of the groups appeared to be doing anything to educate, much less mobilize, their members to stop the 151 proposed coal plants.

Perhaps I shouldn't have been so surprised. The mainstream environmental movement embraces an A-to-Z array of concerns, including endangered species, fisheries, habitat preservation, pesticides, product safety, toxic wastes, and dozens more. The movement might be described as a collection of silos, each silo representing one specialized issue. There is the Arctic Wildlife Refuge silo, the rainforest silo, the acid rain silo, the marine mammals silo, and so forth. In order to be effective, the large environmental groups tend to divide up issues among themselves; internally, their staffs tended to specialize further. Perhaps it was not realistic to think that the entire movement would ever channel its mobilizing energies into a single campaign.

On the other hand, climate pervades and even defines all other aspects of nature, and it's hard to imagine an environmental gain that couldn't be undone by global warming. For example, in order to protect an endangered species, one could expend vast efforts securing the protection of a piece of vital habitat. But all those efforts would be rendered moot if global warming radically altered the climate, making it unsuitable for the endangered animal.

Eventually, I did find two national groups that were mobilizing their members on the coal issue. One was the Sierra Club, the other the Rainforest Action Network. Still, in both cases, the coal campaigns had to vie with various other issues competing for the organization's attention. It seemed strange to me that of all the national environmental groups, not one was focused exclusively on stopping coal. If global warming was the greatest threat to the future of the planet, and if stopping coal was 80 percent of the solution to global warming, then such a highly focused group seemed fully warranted.

Obviously, I was in no position to conjure a new environmental group out of thin air. But I had to do *something*, if only to avoid a sense of utter powerlessness. I decided to create a simple one-page Web site that would give people a capsule description of Hansen's proposal for a moratorium on new coal plants, provide links to news stories and research on the coal, and link people directly to activist campaigns. The whole effort took just a few hours. I kept things as simple as possible, including a straightforward banner headline: "Coal Moratorium Now!" By midday the Web site was completed, and I sat back to admire my work.

"Crude," I thought, "but not a disaster."

Next, I decided to dig deeper into what this new coal boom was all about, wondering why coal was still such a big part of the U.S. energy mix. I soon learned that the existing fleet of about 600 coal plants, many of them dating to the Eisenhower administration, provides about half of the electricity used in the United States. Building a new coal plant, let alone 151, is a vast and expensive undertaking. Consider the dimensions of the typical coal plant, including an immense boiler housed in a twelve-story-tall building and an 800-foot smokestack visible from a distance of fifteen or twenty miles. Writers attempting to describe such construction projects often strain for metaphors: oceans of concrete, forests of steel girders. But no description can quite prepare you for the experience of coming to one of these plants in person, especially during the construction phase, when a workforce of several thousand, housed in its own temporary trailer city, works round the clock on a project whose price tag runs into the billions of dollars.

Once running, a single 500-megawatt plant can burn its way through a 125-car trainload of coal in two days. During combustion, each carbon atom in the coal combines with two oxygen atoms, creating a quantity of carbon dioxide that weighs approximately twice as much as the original train. To offset the carbon dioxide produced by a single coal plant, 850,000 SUV drivers would have to switch to Priuses. Even that comparison understates the consequences of a new power plant, since a car lasts about a decade, while a typical coal plant will continue to spew climate-torquing gases for sixty years or more.

It was easy to see why James Hansen was alarmed by the proposals for 151 new coal plants. Once built, they would become part of the energy infrastructure and would be almost impossible to dismantle, destroying any hopes that global warming might be prevented.

It wasn't supposed to be this way. Coal was the fuel of the past, especially the smoky nineteenth century when fossil fuels replaced animal and waterpower in English mill towns, propelling England as the first country to enter the Industrial Revolution. America and Germany, both well endowed with coal, had followed England's pathway. It was a mixed history. Cities became unhealthy places. Workers consigned to mine work, including children, lived truncated, impoverished lives.

Over time, the use of coal shifted away from everyday uses such as home heating. Instead, it became used primarily for generating electricity. Automation pushed coal production steadily westward, away from the underground mines of Appalachia and toward large strip mines in the Midwest and the West. After the environmental movement in the 1960s forced a recognition that acid rain caused by the sulfur in coal was

ruining forests up and down the Eastern Seaboard, new legislation accelerated the move toward lower-sulfur western coal.

After World War II, electric utilities continued building coal-
fired power plants, but when the federal government changed its
rules in 1992 to encourage the burning of natural gas in power
plants, construction of coal plants virtually ceased. Then in
2000 a jump in the price of natural gas caused the pendulum
to begin swinging back toward coal, as did a friendly shove
from the newly arriving administration of George W. Bush.
Within months of Bush's inauguration in 2001, Vice President
Dick Cheney convened a secretive energy task force, among
the aims of which was to revive the building of coal plants.

The *Washington Post* uncovered a typical piece of business
for the energy task force: In February 2001 Jack N. Gerard, a top
official with the National Mining Association, had a meeting
at the offices of Cheney's staff with task force director Andrew
Lundquist and other staffers. Gerard urged the administration
to put the industry-friendly Department of Energy, rather than
the Environmental Protection Agency, in charge of global warming policy. The administration adopted the recommendation,
scuttling chances for greenhouse gas regulation.

With oilmen Bush and Cheney in charge, energy companies
saw an opportunity to get as much accomplished as possible.
Among the recommendations of the task force was that 1,300 to
1,800 new power plants would be built in the United States by
2020, with an emphasis on new coal-fired plants. Now it was six
years later, and according to the list compiled by Erik Shuster,
151 coal plants were in various states of planning, permitting,
and construction. The list showed coal plants on the drawing
board in thirty-eight states. I was curious to know more. Where
was each proposed plant located? What was the exact status

of each? After I had created the Web site highlighting James Hansen's call for a moratorium on new coal plants, it occurred to me that a useful next step would be to compile a brief status report on each proposed plant and add the information to the Web site.

To find out the status of the plants, it seemed that the best way to proceed was to call around the country and talk to the grassroots groups that tend to do so much of the heavy lifting on environmental issues. I picked up the phone and called Mark Trechock at the Dakota Resource Council (DRC), a farmer/rancher organization located in my hometown of Dickinson, North Dakota. Growing up in that part of southwestern North Dakota, I had often seen the black smoke spewing from a local coal-fired briquette factory, and during the summers I had worked in the shadows of the immense draglines that mine the coal. After college, I'd taken a job as a community organizer for DRC before moving on to other endeavors. But I'd stayed in touch with the group, and Trechock was a good friend. In response to my questions, he quickly updated me on coal projects in North Dakota and suggested that an even quicker way to do my research would be to join a computer mailing list called No New Coal Plants, an online forum that had become a favorite gathering spot for anticoal activists.

"Send an e-mail to Mary Jo Stueve at South Dakota Clean Water Action," he said. "She'll help you get on the list."

■

THREE

Inside the Swarm

■

ON A CHILLY NIGHT in February 2007, a criminal justice consultant named Nancy LaPlaca sat on a bare bench under the
bright lights of the Denver County Jail. Four other women sat
beside her, two arrested for public inebriation, a third brought
in on suspicion of crack possession, the last for driving while
intoxicated. In her day job, LaPlaca had seen many such rooms.
But now she was on the wrong side of the bars.

LaPlaca had begun the evening at the Denver Marriott,
relaxing in the hotel bar with friends after the close of a small
conference that she and her group, Coloradoans for Clean
Energy, had organized for activists from across the country
who are opposing new coal-fired power plants. Next to her
chair she had carefully placed her "NO NEW COAL PLANTS"
sign so that it faced the wall, after a request to do so from the
hotel manager. A utility industry conference was taking place
in the same building, and the manager was eager to avoid
offending the executives and engineers in attendance. But as
LaPlaca prepared to leave, she briefly turned her sign so that
it was visible to the bar.

"Suddenly," she later recalled, "there was this 250-pound policeman in my face demanding to talk with me privately. I told him that whatever he had to say, he could say in front of my friends. And that's when he grabbed me."

LaPlaca told me her story over the phone as she prepared to face a judge on charges of trespass and disorderly conduct. I had found her through the No New Coal Plants listserve that Mark Trechock had recommended to me.

This useful watering hole had been initiated in April 2006 by Philadelphia organizer Mike Ewall. Ewall founded the group Energy Justice Network in 1999 and has organized listserves on issues ranging from tire incinerators to nuclear power. Whatever the topic, the elements of each listserve are identical: messages from any member are forwarded to the entire group, responses may be directed back to either the group or the original author, and archives of group messages are kept on the Energy Justice Network Web site.

For the first few months, messages among No New Coal Plants participants were few and far between. But by midsummer 2006, Ewall had recruited several dozen members, and the listserve had taken on a life of its own. Over the next year, it grew to include 140 people, a membership that was diverse as well as far flung. A few members, such as Matt Leonard of Rainforest Action Network in San Francisco and Ted Glick of the U.S. Climate Emergency Council in Takoma Park, Maryland, were on staff at national environmental groups. Most, however, were involved with small, locally based, mainly rural groups. Typical among these was Greg Howard, an attorney with the nonprofit Appalachian Citizens Law Center, a law firm in Prestonsburg, Kentucky, that represents miners suffering from black lung disease; Mano Andrews of the Western Shoshone

Defense Project in Nevada and the Save the Peaks Coalition in Arizona; and Leslie Glustrom, a biochemist in Boulder, Colorado, opposing Xcel Energy's Comanche 3 coal plant.

As I became better acquainted with participants in the No New Coal Plants listserve, it became clear why the solidarity and shared resources mattered so much. For those living in areas that were already heavily affected by mines and power plants, the struggle was not about the future of the planet. They were fighting for their homes, livelihoods, and health— or even all three at once. One such person was Elisa Young, who was battling to save the farm in Meigs County, Ohio, that her ancestor George Roush had received in compensation for his service in the Revolutionary War. There were already four coal-fired power plants within eyeshot of Young's house, and five more plants were planned for the area. Her group, Meigs Citizens Action Now!, was dealing with the daily nuisances and hassles of existing coal development—blasts, noise, toxic emissions, truck traffic, coal plant waste, contaminated water— while simultaneously working to stop new plants, mines, and waste disposal sites.

To accomplish the latter, Young traveled the state, attempting to persuade cities that were slated to be purchasers of the power from a 960-megawatt coal plant being proposed by American Municipal Power to be wary about the economic consequences of signing on. Like a latter-day Virgil guiding Dante through the circles of Hell, Young also made time to show visitors the various sorts of devastation inflicted on Meigs County: strip mines, coal conveyor systems, haul roads, transmission lines, waste pits, all of which had been affixed to a once bucolic setting of wooded hills, country church graveyards, cornfields, and cattle pastures. Six of Young's neighbors had already died of cancer, and Young herself was being treated for a precancerous condition.

Another member of the listserve who was no stranger to coal development was Indiana photographer John Blair, whose group, Valley Watch, monitored developments along the heavily polluted Ohio River industrial nexus conjoining the states of Indiana, Kentucky, and Illinois. Valley Watch was thirty years old, and many members of the listserve leaned on Blair's long experience. A typical newcomer to environmental activism was Tom Karas, a contractor who built log cabins in northern Michigan. Karas knew his community intimately, and he had already made tremendous progress in mobilizing local citizens in opposition to a project slated for his county known as the Wolverine Clean Energy Venture.

For all participants in No New Coal Plants, the listserve provided a variety of support: research assistance, clipping service, and watercooler. Postings announced conference calls, floated ideas for group projects, celebrated victories.

"This is hard work, with low pay and lots of frustrations along the way," Alan Muller told me. Muller was a former chemical engineer who now served as the one-man staff for Green Delaware. He said, "I can't stress enough the encouragement factor as a main value [of the listserve]."

In some ways the No New Coal Plants listserve actually fit the profile of a single-issue environmental group, if "group" is the right word for an entity with no office, no board of directors, no letterhead, no bank account, no organizational structure. But the term "swarm" would better reflect the anarchic quality not just of the listserve itself but of the movement it represented.

As fighting forces, swarms both preceded and eventually vanquished the orthogonal ranks of legionnaires that forged the

Roman Empire. In a swarm, the emphasis is not on discipline, experience, and orderliness but rather on fighting spirit and individual initiative. Swarms are known for their tactical flexibility, sometimes using guerrilla-style harassment, as did the farmers who routed the British at Lexington and Concord, other times prevailing with overwhelming numbers in the manner of the Arapaho, Lakota, and Northern Cheyenne fighters who overran the U.S. Seventh Cavalry at the Little Bighorn.

The contrast between No New Coal Plants and Big Coal was obvious, but the contrast between such low-profile, decentralized entities and the large national groups typically identified with the environmental movement was equally striking. Typically based in Washington, D.C., or New York and sporting annual budgets in the tens of millions of dollars, these "Big Green" groups, not unlike the corporate and governmental entities they oppose, are hierarchical, highly organized, and reliant on trained and seasoned attorneys, scientific experts, and lobbyists. Yet the "Twigs," a name some small-scale activists used to distinguish themselves from Big Green, had lately taken more militant positions on key aspects of the global warming controversy.

By the time I first began following the anti-coal swarm in the spring of 2007, the difference between the grassroots groups and Big Green had blossomed into a full-blown argument over a pressing issue facing the movement: whether to support a new technology with the ungainly acronym IGCC, for integrated gasification combined cycle.

Rather than create electricity by burning coal, IGCC plants first convert coal into syngas, a mixture of carbon monoxide and hydrogen, then burn the gas. The technology for coal gasification is not new. It was initially used to power the German air force during World War II. More recently, the apartheid regime in South Africa, isolated economically from the rest of the world, had used the technology to supply some of its fuel needs. The use of gasification for electrical generation is relatively recent. Four such plants operate in Europe and the United States, all built with government subsidies. Because it involves converting solid fuel into gas prior to combustion, IGCC technology is better suited to capturing waste products than conventional combustion technology. As much as 88 percent of the coal's carbon dioxide can be captured in an IGCC plant, along with 99 percent of its sulfur oxides and particulates and 95 percent of its mercury. Once the carbon dioxide has been removed from the exhaust stream, it can be liquefied under pressure and injected into deep underground formations. In the spring of 2007, over a dozen IGCC plants were under development in the United States. Leading the pack was Eurora Group's Cash Creek facility, slated to begin operating in Kentucky as early as 2011.

For Appalachian groups whose greatest concern was the destructive mining practice known as mountaintop removal, the fact that IGCC plants would still entail the destructive mining of coal was already a deal breaker. Other grassroots groups had additional concerns about the technology, not trusting that carbon capture and storage could be safely carried out, or believing that the entire enterprise was something of a fig leaf allowing coal companies to continue doing business as usual.

Four prominent groups did support IGCC—the Natural Resources Defense Council, the Environmental Defense Fund, the National Wildlife Federation, and the Clean Air Task Force. Underlying the decision of these groups to work with the coal industry in building the new plants was a brutal calculation by experienced leaders of the larger groups, most prominently

David Hawkins, director of the Climate Center at the Natural Resources Defense Council. Hawkins was one of the most senior figures in the environmental movement, having joined NRDC in 1971. He told other environmentalists they should find ways to leverage the political strength of the coal industry rather than continually hoping they could defeat it. On a visit to Australia, he told journalist Bob Burton, "What we are exploring is whether the political power that is represented by the fossil energy industry can actually be used to move the process forward rather than have them in their traditional role of opposing action."

In April 2007 Hawkins told the Senate's Energy and Natural Resources Committee that "we will almost certainly continue using large amounts of coal in the U.S. and globally in the coming decades." For that reason, he concluded that "it is imperative that we act now to deploy [carbon capture and storage] systems."

A key objection to IGCC involved the efficacy of pumping carbon dioxide underground for indefinite storage. While such pumping had been done to facilitate oil extraction, it had never been attempted at the immense scale that would be required to render the coal industry climate-friendly. According to a study by engineers at Massachusetts Institute of Technology, capturing and compressing just 60 percent of the carbon dioxide produced by U.S. coal-fired power plants would require a new pipeline network big enough to move 20 million barrels of liquefied carbon dioxide each day from power plants to suitable underground storage sites, a volume equal to all the oil piped daily throughout the country. The Department of Energy estimated that by the end of the century, the amount of liquified carbon dioxide needing to be permanently sequestered would be enough to fill Lake Erie twice over or cover the entire state of Utah with a blanket of liquified carbon dioxide 14 feet thick. Storage sites would have to be honestly administered, closely monitored, and tightly sealed. The demanding technical requirements led journalist Jeff Goodell to write that "the notion of coal as the solution to America's energy problems is a technological fantasy on par with the dream of a manned mission to Mars."

A more straightforward concern about IGCC was its economic feasibility. The cost of building such plants was expected to be around 40 percent higher than conventional coal plants. And the cost of operating them would also be higher, since huge amounts of power are needed to separate and liquefy carbon dioxide, then pipe and pump it underground. In all, each plant would have to burn about 25 percent more coal to generate the same amount of electricity for market. Once those expenses were totaled up, this way of using coal seemed headed toward being more costly than electricity generated by solar or wind power.

During the spring of 2007, members of the No New Coal Plants listserve used the network to develop a rapidly growing information base on the projected costs of IGCC. Among those urging research into the costs of IGCC, the most vocal was Carol Overland, an attorney based in Redwing, Minnesota. After working as a truck driver for over a decade, Overland sold her house in the early 1990s to finance a law degree from William Mitchell College of Law in St. Paul, Minnesota. She went to work representing small towns and local groups in transmission-line permitting and other utility-related cases. As a girl, she had played "power engineering office" on a desk made from a red crate, imitating her father, a mechanical engineer who had designed power plants for Great River Energy and other utilities.

Now that childhood game had turned into a career represented by floor-to-ceiling shelves constructed from two-by-fours and filled with power company feasibility studies.

Overland was one of the earliest participants on the No New Coal Plants list and clearly one of the brightest. She had a talent for exposing the financial weak spots of proposed power plants, and she coached others on the list: "If you want to kill a power project, focus on economics."

Overland was applying that advice to the Mesaba Energy Project, a massive IGCC plant being proposed for Bovey, Minnesota, by independent power generator Excelsior Energy. The plant would use coal shipped by rail from Wyoming's Powder River Basin. The coal would be converted to gas and then the gas burned to make electricity, which would be sold to the customers of Minnesota utility Xcel Energy.

For all the claims that Mesaba was a technological step forward, the real creativity of the project seemed to lie in Excelsior Energy's ability to attract government subsidies. Like a confidence man playing a wealthy widow for a big score, the promoters of Mesaba, led by husband-and-wife team Tom Micheletti and Julie Jorgensen, both former employees of Xcel Energy, planned to leverage small grants from the body politic into bigger ones. This led to the choice of a location for the project: the far northeastern part of Minnesota known as the Iron Range. The Iron Range lacked the geology needed for storing liquefied carbon dioxide, but as a region left economically depressed after a century of boom-and-bust iron extraction, it contained something more valuable to Mesaba's developers: state business development subsidies.

In 2002, the Micheletti/Jorgensen team picked up their first $1.5 million grant from Iron Range Resources, a state development agency funded by taxes on taconite mines. The next year, the developers secured an additional $8 million from Iron Range Resources, a $10 million grant from Minnesota's Renewable Development Fund, and a $36 million grant from the federal Department of Energy.

The big money remained to be secured: a federal loan guarantee of up to $1.6 billion and federal tax credits of up to $130 million. In the summer of 2005, the developers decided to change their preferred location from an abandoned mine site near Hoyt Lakes to a scenic area of lakes, forest, and wetlands a hundred miles to the west in Itasca County. The move prompted a frenzy of organizing, as local citizens met in living rooms and public halls to share information and hear speakers, including project sponsors and project critics. They formed Citizens Against the Mesaba Project (CAMP), set up a Web site, and began networking with other grassroots groups around the state.

As Overland watched the Mesaba Project unfold, what galled her most was how the developers had managed to pass it off as a "green" project, not only to members of Minnesota's political establishment but also to major environmental groups in the state. Again and again, Overland pointed out that use of the new IGCC technology in this instance was pointless, since Minnesota lacked the type of geological formations needed for pumping carbon dioxide underground. Many grassroots environmentalists in Minnesota had shifted to opposing the plant, but the matchup remained an improbable one: Overland with her fruit crates against a number of well-connected members of the Minnesota political establishment.

Yet Minnesota is famous as a state of underdogs and mavericks. It's the home of Jesse Ventura, the professional wrestler who became governor, and Al Franken, the comedian turned U.S. senator. There's a sort of Scandinavian puckishness afoot that likes to tweak pretensions and level the field. From conversations off the official record, Overland knew that not everyone inside the regulatory agencies charged with reviewing the Mesaba case was delighted with the plan. They weren't willing to front the argument—that would still be up to the activists—but at least the arguments of Overland and other opponents would get a proper hearing.

To bolster Overland's case, other No New Coal Plants participants supplied her with internal reports on coal prepared by Wall Street investment banks and with feasibility studies performed in other states. These showed mounting evidence that IGCC might not be the wonder technology that its proponents seemed to think. Essentially, an IGCC plant was a refinery joined at the hip with an electricity-generating plant. That posed a problem whenever one or the other system was not working properly. Refineries in particular tend to be fussy and complex, requiring constant adjustment of pressures, temperatures, and catalysts. This meant that a factor often touted in favor of coal—its baseload reliability, especially compared to solar and wind power—could not necessarily be assumed. Moreover, whenever an IGCC plant shuts down, a long restart period is necessary, during which emission levels are typically far higher than the usual specification for the plant.

On top of the cost overruns typically associated with new technologies, the planners for Mesaba were confronting an industry-wide escalation in building costs. Rapid economic growth in China and elsewhere was putting pressure on materials such as concrete and steel. Skilled workers were in short supply. Engineering costs exceeded expectations. While the

U.S. Department of Energy had originally placed the cost of Mesaba Unit 1 at $1.18 billion, by May 2006 that number had nearly doubled to $2.2 billion, not including necessary transmission line upgrades or the needed infrastructure for carbon capture, transportation to a location with suitable geology for carbon sequestration, underground injection, and long-term monitoring.

The more information Overland received, the more she became convinced that an aggressive assault on the cost estimates for Mesaba was the key to derailing the project. In order to build the plant, Excelsior Energy needed the state of Minnesota to approve a power purchase agreement (PPA) between Excelsior and Xcel. In a brief to the Minnesota Public Utility Commission, Overland maintained that Mesaba should not receive the PPA because it did not qualify as a "least cost project" under Minnesota's statutes; given the revised cost projections, Mesaba's electricity wouldn't be as cheap as alternative sources. Having submitted her briefs in quadruplicate, she hunkered down to wait for the regulators to make their first big decision.

In April 2007 the decision was announced. Agreeing with Overland and Citizens Against the Mesaba Project, a panel of administrative law judges recommended to the Minnesota Public Utilities Commission that the PPA be denied on economic grounds.

"Dead, dead, dead!" a jubilant Overland told the *Star Tribune*. "It was on life support before. The plug has been pulled and we're waiting for the inevitable."

Mesaba wasn't actually dead yet. Even a year later, the project's backers continued to pursue subsidies and permits. But the aura of inevitability that had once surrounded the project was gone, and now the sponsors were on the defensive. Within the anti-coal movement, the victory, however tentative, was regarded as highly significant. If a project with so much backing could be successfully challenged, perhaps projects elsewhere were more vulnerable than had previously been assumed.

■

FOUR

But We'll Freeze in the Dark!

■

Unless we tell our politicians to ignore Al Gore's scam,
we'll all freeze in the dark.   —WILL OFFENSICHT

LIKE CAROL OVERLAND, EVERY participant in the No New
Coal Plants listserve was out to stop coal-fired power plants.
But was this a responsible position to take? Electricity, after
all, is the lifeblood of modern society, the effect of its absence
a rapid descent into chaos.

Opponents of coal scoffed at the idea that a moratorium on
new coal-fired plants would pose any threat to the country's
energy security. They pointed to the existence of ample reserves
of electrical capacity and to studies showing massive untapped
potential for expansion of wind, solar, and geothermal resources,
at costs competitive with coal plants. They also pointed to the
large gains in energy efficiency that were also available, at an
even lower outlay.

Alan Muller of Green Delaware observed that costs for
renewables like wind were dropping while coal plant costs
were rising quickly. In a fair matchup with renewables, he

believed that coal would lose. The key step, therefore, was to create regulatory procedures that forced coal plant proposals into one-on-one cost competition with alternatives. If such matchups could be made a regular step in the consideration of new plants, Muller was confident that coal would lose.

In 2007 Muller got a chance to test his hypothesis, as Delaware put into effect a new process for judging utility expansion proposals known as integrated resource planning, or IRP, and at the same time began evaluating competing proposals for new power supplies. One proposal, from NRG Energy, was a coal-fired power plant known as Indian River. A competing proposal, from Bluewater Wind, involved offshore wind farms located about eleven miles from the coast, and backup power provided by natural gas turbines.

In Delaware, a public opinion survey by the University of Delaware showed strong support for wind and strong opposition to increased coal generation. But Muller felt that despite such sentiments the state was committed to its analytical process and would not choose the Bluewater alternative unless the cost data strongly supported that option. Ensuring that the numbers being provided by the bidders were valid was impossible to verify, since both NRG and Bluewater were seeking to prevent public disclosure of their respective bids. Muller suspected that NRG was supplying Delaware officials with low-ball figures, and he appealed to members of the No New Coal Plants listserve in other states for cost studies from other pending coal plant cases.

In response to Muller's appeals, data poured into Delaware from dozens of activists across the country: Colorado, Minnesota, and elsewhere. The most timely information came from Carol Overland, whose work in the Mesaba case had unearthed a trove of data showing dramatic increases in the costs of IGCC. Overland flew to Delaware and met with state officials to present the numbers. When the dust had settled, Delaware announced that the Bluewater Wind proposal had been chosen. "Carol's numbers drove the nail in the NRG coffin," said Muller.

Across the country, others were finding the cost of wind power increasingly favorable compared to the cost of new coal power. An analysis by the investment banking company Lazard Ltd. found the cost of generating electricity from coal to be 7.4 to 13.5 cents per kilowatt-hour (the high end included carbon capture and storage) while the cost from wind was estimated to be 4.4 to 9.1 cents per kilowatt-hour. A study released by the California Energy Commission estimated a cost range for coal of 10.6 to 17.3 cents per kilowatt-hour, compared with 8.9 cents per kilowatt-hour for wind. Mass production of wind turbines promised to lower costs even further. By the end of 2007, worldwide wind capacity had exceeded 93,000 megawatts and was on course to nearly double in three more years.

The U.S. Department of Energy (DOE) released a study showing that wind could supply 20 percent of the country's electricity needs by 2030. Under this scenario, wind would displace 50 percent of electric utility natural gas consumption and 18 percent of coal consumption, at costs ranging from 6 to 10 cents per kilowatt-hour, including the cost of connecting the wind into the grid. About a sixth of this power would be produced by offshore wind farms like the Bluewater proposal, bringing power to populated urban centers. Nor would the demands placed on U.S. manufacturing capacity be excessive. In the peak year of the buildout, the DOE study called for 16,000 megawatts of new capacity, an amount comparable to the amount of new gas turbine capacity installed in the United States in 2005.

Of course, in order to create the utility demand that would bring wind farms into actual existence, coal plants needed to be canceled. For that reason, Delaware's Indian River decision was particularly significant, because it shattered the conventional wisdom that coal is the lowest-cost way to provide power.

Wind was just one of several technologies that offered an alternative to new coal plants. Another was solar thermal, which energy analyst Joe Romm called "the solar power you don't hear about." In this surprisingly straightforward way of generating electricity, acres of mirrors heat pipes containing water or molten salt. The heated fluid in turn drives turbines to create electricity.

Under prodding by the California Energy Commission (CEC), solar thermal was rapidly moving into a position to become a major supplier of the electric grid for that state. The CEC liked the technology because its costs were estimated to be 27 percent lower than new coal plants with carbon capture and storage—12.7 cents per kilowatt-hour for power from a solar thermal plant versus 17.3 cents per kilowatt-hour for power from a coal plant equipped with carbon capture-and-storage technology.

During 2007, numerous solar thermal plants were moving forward, not only in the western United States but also in Europe. Several of the plants included on-site thermal storage, a feature that makes solar thermal a reliable source of baseload power. For example, in Spain, the Andasol 1 plant included large tanks containing tons of molten salts that absorbed heat during sunny periods and released it to generate power during cloudy periods or nighttime. The result was 7.5 hours of thermal storage and the ability to generate power for nearly twenty-four hours per day.

According to David Mills, chairman of solar thermal pioneer Ausra, a rectangle of land in the sunny southwestern United States measuring about ninety-five miles on each side, if devoted to solar thermal installations, could fully supply the U.S. electric grid. With favorable locations for solar thermal plants, Morocco could similarly supply power to Europe, as could the Gobi Desert to China. The necessary amount of land, while sizable, is about the same as the amount disturbed by coal mines, which are far more destructive. It would be just one-sixth of the area devoted to lawns, one-fifteenth of the area once devoted to raising feed for horses, and one-thirtieth of the area devoted to parks, wilderness, and wildlife refuges.

Further evidence that economically attractive alternatives to coal could be developed was contained in the report *The Future of Geothermal Energy*, penned by an eighteen-author team at the Massachusetts Institute of Technology and released in 2006. The report focused on the potential for enhanced geothermal power, a method for exploiting the hot dry-rock resource that exists nearly everywhere at depths of three to ten kilometers. To provide steam for an enhanced geothermal plant, deep wells are drilled, followed by injection of cold water to produce a network of cracks in the rock. Water is then pumped into the fractured rock and harvested as steam for generating power. According to the MIT study, the necessary step toward developing enhanced geothermal power is a government-financed research and development program to refine today's deep-well drilling technology. The study estimated that 100 gigawatts of enhanced geothermal plants could be built by 2050, an amount sufficient to replace about a third of today's coal plants, at a cost cheaper than building new coal plants.

FIGURE 2 PER CAPITA ELECTRICITY USAGE IN CALIFORNIA AND THE UNITED STATES, 1960-2005

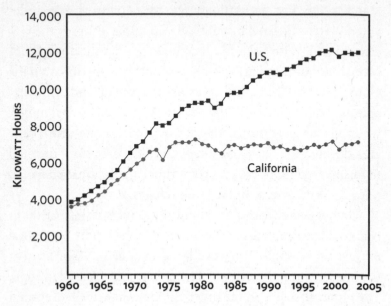

Beyond wind, solar, and geothermal power, a way of supplying energy needs existed that was even more competitive and plentiful: efficiency and conservation measures. Some environmentalists suggested that using the term "negawatts" was the best way to convey that energy savings weren't just a matter of changing behavior by consumers, but rather could be proactively effected through utility investments and tougher standards for buildings and appliances.

To anyone who questioned the potential size, cost, or effectiveness of negawatts, the answer could be summed up in a single word: California. As shown in figure 2, electricity consumption patterns in California were the same as those in the rest of the United States until the early 1970s. But then something happened. Beginning in 1973 and continuing for the following three decades, California's electricity usage flattened out, while that of the rest of the country continued to rise another 50 percent.

The difference was astonishing. Sixty large coal plants that otherwise would have been necessary were not built in California. The main reason for California's lower energy usage was a bevy of state-mandated efficiency improvements that were largely invisible to the average citizen of the state. The flattening of per capita usage had been named the Rosenfeld Effect in honor of the Lawrence Berkeley Laboratory physicist responsible for many of the innovations, Art Rosenfeld.

Rosenfeld had been the last student of Enrico Fermi, one of the leading physicists behind the Manhattan Project. He was forty-six, with an extensive career in basic physics already behind him, when his moment of destiny arrived with the OPEC oil embargo of 1973. The embargo created a crisis across the United States, as lines of cars formed at gas pumps and a sense of panic filled the air. Rosenfeld's response was to bring a collection of experts in the fields of energy, utilities, transportation, and building design together for a month-long brainstorming session at Princeton University. One of the surprising findings of the meeting was that buildings alone account for two-thirds of the electricity used in the United States each year.

Recalling the watershed conference, Rosenfeld later said, "We realized we had found one of the world's largest oil and gas fields. The energy was buried, in effect, in the buildings of our cities, the vehicles on our roads, and the machines in our factories. A few of us began to suspect that the knowledge we gained during that month would change our lives."

Back in Berkeley, Rosenfeld founded the Center for Building Science, which over the next two decades developed a broad range of energy efficient technologies, including the electronic ballasts that led to compact fluorescent lamps, and a window material known as "smart glass" that blocks heat while allowing light to pass through.

Not content merely to develop such ideas, Rosenfeld pushed them into the California state policy arena. Luckily, the governor of California, Jerry Brown, reveled in new ideas. Under Brown's watch, California developed a bureaucratic structure to implement energy conservation. In addition to developing hardware, much of Rosenfeld's work had to do with developing policy mechanisms to make the electricity market "smarter" so that price signals could translate more effectively into conservation. For example, he pushed for time-of-day pricing, so that consumers and businesses that shifted their energy use to evening hours could benefit from lower rates and power company "peaks" could be smoothed off, eliminating the need for power plants. Another idea was smart meters, which could receive electronic signals offering lower prices for cutting back at critical times.

Each such innovation may seem trivial until you consider the size of the markets involved. There are about a hundred million refrigerators in the United States—maybe more. In the early 1970s refrigerators were lightweight and noisy. Rosenfeld and crew upped the efficiency of refrigerator motors from 30 percent to 90 percent and added insulation. The result was a machine that used a quarter of the electricity that it previously required and saved its owner $200 or more per year. Due to refrigerator improvements alone, a hundred large coal plants that would have been required were no longer needed.

This was the best answer to "If you block this coal plant, we'll have rolling blackouts and the lights will go out." To build a coal plant requires eight or more years of planning and construction. But efficiency measures can be implemented much faster.

Meanwhile, the reverse is true. If coal plants are built, utilities develop a powerful incentive to run those plants and have no reason to invest in alternative ways of meeting their customers' need for electricity; indeed, when utilities have excess capacity, they may even discourage rather than facilitate conservation measures. So energy efficiency and stopping coal plants are two efforts that work hand in hand.

Perhaps the most astonishing thing about the California energy revolution was how cheap it was. Innovations such as low-flow showerheads or tighter insulation standards for new housing cost 1 or 2 cents per kilowatt-hour, about a tenth of the cost of building a new power plant. Strict energy efficiency standards for refrigerators pushed manufacturers to innovate in ways that actually saved rather than cost money. Rosenfeld's energy-efficient windows, which were enabled by the careful development of a high tech film coating, can reduce a building's energy use by 30 percent. Such windows yield many times more in savings than their initial cost.

Other than the entrenched political power of the coal and utility industries, there was no reason that the innovations Rosenfeld and his team had developed could not be adopted around the country. If that were to happen, it would hugely affect how many coal plants would be built in the future, if any.

Even a slight downward adjustment in projected growth rates is capable of having a dramatic effect on the building of new coal plants, since the expectation for growth is a prereq-uisite not just for utilities to plan new coal plants but also for regulators to approve them and banks to finance them. This became obvious when a bureaucrat named Guy Caruso caused 132 coal plants to disappear with a wave of his magic mouse.

Caruso was the head of the Energy Information Administration (EIA), which in 2007 projected that electricity consumption would grow at the rate of 1.5 percent per year through 2030. But on March 4, 2008, Caruso told Congress that the EIA had decided to adjust that number to 1.1 percent.

A change from 1.5 percent to 1.1 percent annual growth may not sound significant, but by 2030 the lowered growth rate would reduce the projected electricity generation requirements by the equivalent of 132 coal plants, each rated at 500 megawatts. While the EIA administrator does not actually decide which power plants are going to be built—that's done by individual utilities and power authorities, each making its own economic and power growth projections—the EIA projections do set the tone for governmental policy at all levels. So even though 132 coal plants weren't directly canceled by Caruso's scaled-back projection, the revision was a signal to utilities, state agencies, banks, and others involved in the planning and approval process: be careful not to overextend yourself in coal.

This admonition had a historic precedent in the fiscal meltdown of the nuclear industry. During the 1970s and 1980s, many utilities had committed themselves to immensely expensive nuclear plants that required a decade each to plan and build. During that period, costs leaped upward as did interest charges, exhausting and even bankrupting utilities that had once thought nuclear would be "too cheap to meter."

In terms of avoiding expensive overbuilding, alternative ways of supplying power, such as solar, wind, and efficiency investments, enjoyed an advantage. Such technologies could typically be deployed in a year or two. With such short lead times, utilities could control the amount of new capacity more precisely, raising investments during boom times and culling them during recessions.

The combination of slowing growth, new efficiency measures, and emerging renewables provided a promising pathway not just for halting the construction of new coal-fired power plants but for phasing out the existing fleet of plants. That vision was fleshed out in a detailed energy plan released by Google, Inc. Under the Google "Clean Energy 2030" plan, by 2030 the use of coal and oil would end, natural gas usage would be halved, and oil used for cars would decline by 38 percent. The plan would implement the following measures:

- End-use electrical energy efficiency improvements sufficient to reduce demand by 33 percent
- 300 gigawatts (GW) of onshore wind power
- 80 GW of offshore wind power
- 170 GW of photovoltaic power
- 80 GW of solar thermal power
- 15 GW of conventional geothermal power
- 65 GW of enhanced geothermal power

The fact that the plan was developed by one of the world's most respected high-tech companies gave it immediate credibility, as did the high-profile backing of Google CEO Eric Schmidt. On radio and television and at numerous conferences and seminars, Schmidt emphasized that Google's plan could be justified not merely for its environmental benefits but on a cost basis alone. Discussing the plant with the *Wall Street Journal's* Alan Murray, Schmidt said, "I make the argument this way.

You've got to solve a whole bunch of problems. You've got to solve the energy-generation problem, and you've got to solve the transportation problem. So when you add it all up, if you make, in our view, the right assumptions and you invest in the right ways, you end up saving money. That's the thing that was most surprising to me. So the rough numbers are, we need about $3.5 trillion of investment over 22 years, as opposed to over three months, and we generate on a cost basis a savings of $4.4 trillion. If you invest in the right way, you can make money by doing this."

■

What About China?

■

LET'S ASSUME THE UNITED States could phase out its coal plants over the next two decades. Would that be enough to prevent dangerous climate change? Anyone opposing coal on climate grounds quickly runs up against the "China question," typically phrased something like this: "What good will it do to stop coal plants in the U.S.? Aren't they building one every week in China?"

Without a doubt, the scope of the coal boom in China has been breathtaking. According to a frequently cited statistic, the country is building not just one but two midsized coal plants each week, amounting to a yearly increase equivalent to the entire power grid in the United Kingdom. In 2008 annual coal output in China reached 2.76 billion metric tons, more than twice the output of the United States and an astonishing 47 percent of world production.

But there is a limit to how far China's surging coal use can go. Like a cartoon character who runs off a cliff with his legs still moving, China cannot escape the inconvenient reality that its domestic coal reserves were too small to sustain this level of coal consumption. The evidence for that conclusion was laid bare in an analysis of worldwide coal reserves completed in 2007 by the Energy Watch Group (EWG), a private research group initiated by German parliament member Hans-Josef Fell.

The EWG study has an interesting backstory. It was commissioned in the wake of startling revelations about the quality of data on coal reserves in two major European countries, Germany and the United Kingdom. Even though coal had been mined for centuries in Germany and the United Kingdom, neither country seemed able to judge accurately the quantity of mineable coal under its own feet. Germany for years had estimated that it harbored coal reserves in the tens of billions of tons. But in 2004 the country looked more closely at those reserves and concluded that the amount of economically obtainable hard coal amounted not to 23 billion tons but rather to just 183 million tons, a 99 percent reduction. In the same year, the United Kingdom made a similar head-swirling 99 percent adjustment, revising its published reserves from a figure of 45 billion tons reported in 1980 to just 220 million tons.

The revelation that Germany and the United Kingdom had hugely overestimated their coal reserves led the authors of the EWG study to question whether other countries had done the same. After researching the question, they concluded that the "data quality of coal reserves and resources is poor, both on global and national levels."

In general, the reason countries tend to overestimate their coal reserves is not that they don't know the size of the total *resource*—how much coal is physically located in the ground. Unlike oil and gas, coal deposits are not hard to find. Rather, it has to do with the unrealistic assumptions about the amount of coal that can be economically mined. As a rule, recoverable reserves are far smaller than total resources. Most of the easy coal—the high-quality grades lying in relatively thick seams close to the surface—has already been exploited. What remains are deeper, thinner, lower-quality seams, often in inaccessible locations such as beneath roads, rivers, cities, sensitive environmental zones, or industrial facilities. All these factors, alone or in combination, serve to render the bulk of the coal resource uneconomical to mine.

The EWG study found that Germany and the United Kingdom were not alone in wildly overestimating their reserves of coal. In 2004 the "proved recoverable reserves" reported by Botswana dropped from 3.5 billion tons to 40 million tons, another 99 percent reduction. In the United States, a comprehensive study conducted by the National Research Council concluded:

> [I]t is not possible to confirm the often-quoted suggestion that there is a sufficient supply of coal for the next 250 years. A combination of increased rates of production and more detailed analyses that take into account location, quality, recoverability, and transportation issues may substantially reduce the estimated number of years of supply.

Even for the most actively mined coal-producing regions in the United States, published coal reserve figures were turning out to be in serious error. After conducting a detailed assessment of the Gillette coalfield in Wyoming, the source of 37 percent of total United States production, the U.S. Geological Survey concluded that the portion of the coal that can be mined, processed, and marketed at a profit, based on conditions in 2007, was less than half the estimate arrived at by a 2002 study of the same field.*

In the case of China, the Energy Watch Group pointed out that the official figure of 115 billion metric tons of recoverable reserves had not been updated since 1992, "in spite of the fact that about 20 percent of their then-stated reserves have been

---

* For more on constraints to U.S. reserves, see chapters 12 and 14.

produced since then." Figure 3, developed by the Energy Working Group, shows historic and projected coal production in China. According to the EWG projection, the limited size of China's coal reserves will place serious constraints on domestic production in the coming years:

> This scenario demonstrates that the high growth rates of the last years must decrease over the next few years and that China will reach maximum production within the next 5–15 years, probably around 2015. The already produced quantities of about 35 billion tons will rise to 113 billion tons (+ 11 billion tons of lignite) until 2050 and finally end at about 120 billion tons (+ 19 billion tons of lignite) around 2100. The steep rise in production of the past years must be followed by a steep decline after 2020.

In other words, although the pace of China's consumption of its coal is alarming, that pace will be sustained only if China becomes a massive coal importer.

FIGURE 3 HISTORIC AND PROJECTED COAL PRODUCTION IN CHINA

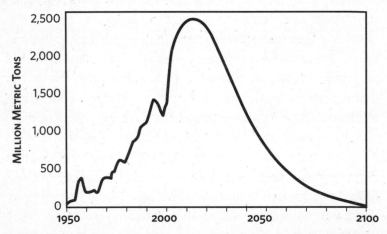

Source: Adapted from *Coal: Resources and Future Production*, Energy Watch Group, EWG-Series 1/2007, March 2007, Figure A-3.

Where might such imports come from? Currently, the runaway leader in coal exports is Australia, with 231 million metric tons of coal moving through the nation's ports in 2006. Yet even if Australia were to sell its exports exclusively to China, it would amount to less than 10 percent of China's annual coal appetite.

Fortunately, China's leadership appears to recognize that the country's coal addiction must be addressed. Driven by a 2005 law that gives incentives such as fixed rate tariffs and carbon credits to renewable-energy companies, expansion of renewable energy has been proceeding at a torrid pace. In 2008 wind power capacity reached 12,000 megawatts; a year later the country was reported to be building six large wind farms, each with capacity of 10,000 to 20,000 megawatts. Similarly, China was moving quickly to become the world's leading manufacturer of photovoltaic cells. In terms of overall investment in renewables, the country was second worldwide (after Germany) and was expected to lead the world by 2009.

Two other aspects of China's coal consumption and energy usage must be kept in mind. First, the per capita contribution of greenhouse gases in China remains far lower than that in the United States—four tons of greenhouse gases per person for China, twenty tons per person for the United States. This difference is due to the country's lower standard of living and to the fact that its consumption of coal, while relatively high, is partially offset by low rates of oil and natural gas consumption. Compared to China's economy, the U.S. economy has far more potential to make efficiency improvements without reducing the basic standard of living.

Second, China's contribution to the greenhouse gases currently in the atmosphere, a reflection of historical consumption, remains far lower than that of the United States not only on a per capita basis but on an absolute basis. Despite having less than a quarter of China's population, the United States continues to bear the biggest responsibility for putting the entire world on course toward serious climate change.

Given the determination of China's leaders to raise the country's standard of living above its historical status of impoverishment, it is inconceivable that China's government would accept limitations on its usage of coal without an equal or greater commitment by the United States and other industrialized countries to do the same. In a very real sense, progress in Chinese policies toward limiting coal usage cannot be expected unless Western anti-coal activists are able to limit their countries' coal consumption, particularly in the United States.

In sum, those who feel that China's rush to build coal plants makes the quest for a U.S. coal moratorium irrelevant have it exactly backward. If we want China to clean up its act, we have no choice but to take the lead.

■

SIX

Hurricane Politics

■

BY THE LATE SPRING of 2007, James Hansen's appeal for a moratorium on new coal plants seemed less hopeless than just a few months before. The back-to-back defeat of proposed coal plants in Minnesota and Delaware had given a shot of confidence to the anti-coal movement. Other political and economic developments also offered promise, including the new embrace of green generation technologies by large venture capital firms, whose support lent business credibility to the general perception that renewables were crossing the magic economic threshold of "cheaper than a coal plant."

Meanwhile, important parts of the political establishment were taking tentative steps toward ending the Bush era of denial and inaction on global warming. A key marker of that shift was an April 2007 decision by the U.S. Supreme Court stating that the U.S. Environmental Protection Agency had both the authority and the responsibility to regulate carbon dioxide and other greenhouse gases. That decision had not yet worked its way into any specific decisions on new power plants, but sooner or later it was sure to become a factor.

In utility boardrooms, the consensus seemed to be growing that some kind of carbon control legislation was inevitable and that this legislation would force utilities to pay a fee or tax for every ton of carbon dioxide they emitted. In South Dakota, groups opposing the Big Stone II power plant had convinced regulators to take such risks into consideration. Similarly, opponents of the Glades Power Plant in Florida pressed that state's Public Service Commission to incorporate the issue into its deliberations on the project.

While the Bush administration remained a fortress of climate change denial, policy was moving forward at lower levels of government. In the Pacific Northwest, Washington governor Christine Gregoire signed a bill that prohibited new coal-fired power plants whose emissions exceeded 1,100 pounds of carbon dioxide emissions per megawatt-hour of electricity generated— the equivalent of a naural gas–fired plant. Since no coal plant could meet the standard without carbon capture technology, the effect of the new law was to create a de facto moratorium on conventional coal-fired power plants in the state. California had already passed a similar bill during the previous year.

In British Columbia, Premier Gordon Campbell delivered a "throne speech" in which he announced a major initiative against global warming by the provincial government. The initiative included a requirement that "all new and existing electricity produced in B.C. will be required to have net zero greenhouse gas emissions by 2016." In 2007 New Zealand began a ten-year moratorium on all new state-owned power plants that used coal, oil, or natural gas as fuel. Ontario's government went even further, committing to a full phasing out of all coal generation in the province by 2014. Premier Dalton McGuinty said, "By 2030 there will be about 1,000 more new coal-fired generating stations built on this planet. There is only one place in the world that is phasing out coal-fired generation and we're doing that right here in Ontario."

A smattering of U.S. utilities was taking voluntary steps to move past coal. In May 2007 Progress Energy, serving approximately 3.1 million customers in the Southeast, announced a two-year moratorium on the construction of new coal-fired power plants. At other utilities, coal plants simply dropped out of long-term plans without public announcement.

As the tide turned against coal, activists felt emboldened to take a harder line. Matt Leonard of Rainforest Action Network, which was focusing its efforts on pressuring Wall Street banks to stop funding coal projects, had been keeping a list of derailed coal projects, and as the list grew by the month Leonard noticed a "ratcheting" phenomenon. He told me, "Whenever activists fighting a coal project in one place are able to get regulators or banks to commit to a certain set of restrictions or conditions, the campaigns against other projects make those conditions the new baseline that must be met or beat. Successes in blocking coal plants are piggybacking from one to the next."

The rising militancy at the grassroots seemed to take some seasoned staff members in national environmental groups by surprise, especially when activists leveled criticisms against an agreement reached between two large environmental groups, the Natural Resources Defense Council (NRDC) and the Environmental Defense Fund (EDF), and an investor group led by the private equity firm KKR, which was in the process of buying the Texas utility TXU. Under the terms of the deal, the new owners of TXU would drop eight of eleven planned new coal plants in Texas; in return, the environmental groups would drop their opposition to the remaining three.

Spokespeople for NRDC and EDF announced the deal to the press as a major success. Many grassroots environmentalists, however, were skeptical. They maintained that TXU would not have bargained away the eight plants if it had believed it would be able to build them; the net effect of the deal was to give up the chance to fight the last three plants. Climate scientists were calling for a full halt on new coal, not a slowdown, they said. If this was the environmental movement's batting average on a good day, it wasn't good enough. A correspondent to *Texas Monthly* wrote: "I feel like I'm in some colonial third world outpost watching helplessly as my fate is being decided by a bunch of rich white guys with Marks-a-Lots in a map room thousands of miles away."

To many grassroots activists, Big Green groups had not grasped how much the growing movement against coal had changed the basic political calculus, mandating a tougher stance toward new coal plants. Before the deal, a wide coalition had marshaled an impressive array of opposition to the plants, culminating with a large rally at the state capitol in Austin and the introduction of bills in the state legislature calling for a statewide moratorium. Now, in the wake of the deal, the push for the moratorium quickly dissipated, though local opposition continued against the three plants that had been allowed to continue.

But whether the TXU deal was shrewd or foolish, one thing it clearly lacked was anything that might inspire and build a mass movement against climate change. In contrast, grassroots activists were beginning to come to the realization that taking a hard-core "no new coal plants" stand was not only a clearer message for rallying support, but also a fully defensible one from the standpoint of cost comparisons.

Debates over when to compromise and when to take a hard line would continue for some time around the country throughout the remainder of 2007. But in one state, Florida, such debates were about to be settled decisively, with the hard-liners winning the argument. The activists who brought the matter to a close were two grassroots environmentalists, Bob and Jan Krasowski. The two had taken advantage of a regulatory provision allowing ordinary citizens to intervene directly in Florida Public Service Commission (PSC) hearings on power plants. In the permitting process for the proposed Glades coal-fired power plant, located on the northwest shore of Lake Okeechobee at the edge of the Florida Everglades, mainstream environmental groups had adopted a complex position. In a memo to the PSC, lawyers for those groups wrote that "[although] there is no need for ... any type of coal plant by FPL [Florida Power and Light], an IGCC plant in Florida can provide electricity at a lower cost than the proposed ... coal plant."

Meanwhile, a similar coalition of Big Green groups, including NRDC, Audubon of Florida, the Florida Wildlife Federation, and the Clean Air Task Force, lobbied on behalf of legislation that would subsidize IGCC coal plants through accelerated recovery of construction costs from utility ratepayers.

To the Krasowskis, the "no ... but" position was a mixed message that mistook the rapidly changing attitudes of Floridians—threatened by both hurricanes and rising sea levels—toward global warming. Convinced that regulators would be receptive to unequivocal assertions by anti-coal forces, the Krasowskis simply demanded that Glades be canceled and replaced with conservation programs like those already implemented in other states.

For support of their position, the Krasowskis sought help from

Alan Muller and Carol Overland. Again, the No New Coal Plants list proved its value as a research service. In the end, it was the Krasowskis' grassroots perspective that prevailed with the Florida PSC in a 4–0 vote that caught most observers off guard.

"We weren't surprised," said Bob Krasowski. "We knew that the commissioners are politically attuned; they have their ear to the ground. And we knew how Florida was leaning. Being in the schools, Jan hears what kids are saying, and that's a pretty good indicator of where their parents are at. As for myself, I constantly hear people in the construction trades talking about how global warming is going to raise insurance rates."

Indeed, awareness among Floridians about global warming was years, if not decades, ahead of the rest of the United States. The reason could be summed up in a single word: hurricanes. Only the oldest people in the state could remember the 1928 Okeechobee Hurricane, which had killed over four thousand people, or the Labor Day Hurricane of 1935, the strongest recorded hurricane ever to strike the United States. But in the decade beginning in 2000, the pace of hurricanes and tropical storms had quickened, and climate change was being blamed. In August 2004 Hurricane Charley made landfall on Cayo Costa with winds of 150 miles per hour before moving onto the mainland. Although modern warning systems meant that the death toll from the storm was far less than that from storms like the Okeechobee or Labor Day hurricanes, property damage was immense, topping $13 billion. Less than a month later, Hurricane Frances arrived, causing another $8 billion in damage, followed immediately by Hurricane Jeanne, which hit the same areas and resulted in billions more in losses. The next year, Hurricane Wilma produced a blackout affecting 98 percent of south Florida, and left in its wake over $20 billion in losses.

Next came Hurricane Katrina, in 2005. Although the storm itself delivered only a glancing blow to the Florida panhandle, the televised images of suffering in Louisiana affected Floridians deeply. A *St. Petersburg Times* poll released in May 2007 reported that 54 percent of state residents believed that global warming had contributed to an increase in the number and severity of hurricanes; 71 percent of those polled said they supported immediate legislative action to cut greenhouse gas emissions.

In July, shortly after the rejection of the Glades Plant, Governor Charlie Crist followed up with a series of executive orders. One order committed Florida to reduce greenhouse gas emissions 10 percent by 2012, 25 percent by 2017, and 40 percent by 2025. It also mandated higher energy efficiency in government buildings and fuel efficiency in state vehicles. A second order established a maximum allowable emission level of greenhouse gases for electric utilities. The standard mandated a reduction of emissions to 2000 levels by 2017, to 1990 levels by 2025, and to 80 percent below 1990 levels by 2050. A final order directed the Florida Public Service Commission to adopt by 2020 a 20 percent Renewable Portfolio Standard that emphasized solar and wind energy.

Crist's dramatic moves showed how fast policy could change when the public reached what Al Gore had described as a "tipping point":

> Sometimes, the political system is like the climate system, in that it's nonlinear. It can seem to change at a snail's pace and then suddenly cross a tipping point beyond which it shifts into a shockingly fast gear. All of a sudden, change that everybody thought was impossible becomes matter of fact. In 1941, it was absurd to think the U.S. could build a thousand airplanes a month to fight the Second World War. By 1943 that was a real small number.

General policy wasn't all that Crist intended to change—he also wanted utilities to start canceling power plants. Shortly after his inauguration in January 2007, Crist dispatched one of his closest aides, former Florida attorney general Chris Kise, to conduct one-on-one talks with utility officials. In meeting after meeting, Kise hammered home the message that the governor wanted no further coal plants to be built in the state. Either utilities could adjust their plans voluntarily or they could wait to see those plans shot down, as the Glades Plant had been, by a Public Service Commission whose members had all been appointed by Crist. At the time, four plants were under consideration.

The first utility to surrender to the governor's demand was the Florida Municipal Power Agency, which in July withdrew its state permit application for the 800-megawatt Taylor Energy Center shortly after the Florida Public Service Commission's rejection of the Glades Power Plant.

Next in the permitting queue was Seminole Electric Power Cooperative, which was seeking to build the 750-megawatt Seminole 3 Generating Station. In August, a month after the cancellaion of the Taylor Energy Center, Florida's Department of Environmental Protection denied a state air permit to the Seminole Station on the grounds that the plant would not minimize environmental and public health impacts and would not serve the public interest.

The third utility, Tampa Electric, suspended the 630-megawatt expansion of its Polk Power Station in October, citing Governor Crist's push to reduce carbon dioxide as the main factor in the decision.

Meanwhile, the sponsors of the fourth project, Orlando Utilities Commission and the Southern Company, had gone ahead and broken ground on the 285-megawatt Stanton Energy Center. But in November, citing concerns about future carbon controls in Florida, company officials announced that they were canceling the project. For a utility to cancel a power project after pouring concrete was highly unusual, an indication of how dramatically the status quo had shifted in Florida.

In May 2008, a year after the *St. Petersburg Times* poll, investigators from Yale University and the University of Miami conducted a new poll. This poll found that 80 percent of the public felt that global warming would cause worse hurricanes during the next fifty years, and 69 percent felt it was somewhat or very likely that rising sea levels would force the abandonment of parts of the Florida coast over the same period. As for which politicians Floridians would trust to tell them the truth about global warming, only Charlie Crist scored a majority (54 percent), while Senators Barack Obama and John McCain, as well as President George Bush, all scored less than 50 percent. Clearly, Crist's aggressive action had boosted his popularity. In fact, 62 percent of those questioned by the pollsters felt the governor should do even more to address global warming. Hurricane politics had changed Florida. But the questions remained: Were the climate politics of Florida a political anomaly that would not be repeated elsewhere? Or was the state merely the first to feel the heat and respond accordingly?

■

SEVEN

Kansas

■

WHILE FLORIDA'S GOVERNOR Charlie Crist deserves accolades for his actions to nix five proposed coal plants in that state, in terms of sheer political courage, no politician in the country compares to Kansas governor Kathleen Sebelius, who never faltered throughout a protracted fight over a three-unit coal plant proposed for the town of Holcomb in the western part of the state. Technically, Rod Bremby, secretary of the Kansas Department of Health and Environment, made the decision to deny Sunflower Electric Cooperative the necessary air permit. But it was Sebelius who four times used her veto power to quash legislation that would have overturned Bremby's decision.

As the most hard-fought battle over coal anywhere in the United States, the Sunflower case was widely seen as a make-or-break moment for the anti-coal struggle. It was also legally significant as the first case in which a state official explicitly based a coal plant decision on the U.S. Supreme Court's 2007 ruling in *Massachusetts v. EPA* that carbon dioxide should be regulated under the Clean Air Act. What surprised many observers was the willingness of Sebelius, once she had made up her mind, to defy the coal lobby in one of the most politically conservative states in the country. What made Sebelius's actions all the more surprising was that the governor had staked her career on a reputation for pragmatism, not for hard-line positions.

As described by Thomas Frank in his book *What's the Matter with Kansas?* politics in the state have been drifting to the right for decades. Republicans now outnumbered Democrats by nearly two to one, and Republican Party policies were generally controlled by hard-core ideologues. Despite the conservative climate, Sebelius had won her first governorship election in 2002 by a 53 to 45 percent margin. She moved quickly to establish a reputation for nondoctrinaire competence, erasing a $1.1 billion state deficit without raising taxes or cutting education. Kansans rewarded Sebelius by reelecting her in 2006 in a 58 to 40 percent landslide. She continued to build her popularity by courting Kansans who felt marginalized by the Republican Party's hard turn to the right, recruiting a number of centrist Republicans into her administration. By 2007, when the confrontation over the Sunflower plant came to a head, Sebelius was riding on approval ratings of 60 percent among Republicans and 75 percent among Democrats, according to polls by Kansas stations KWCH and KCTV. But that popularity had no guarantee of lasting in a state where most people defined themselves as conservative and where conservative media outlets dominated the airwaves. Already, Sebelius had risked the wrath of pro-life advocates by vetoing multiple pieces of anti-abortion legislation and had angered the gun lobby by vetoing concealed weapons laws. In both cases she had managed to walk away with her high approval ratings intact, even among abortion opponents and gun owners.

In deciding to confront the Sunflower project, Sebelius was again taking a political gamble that presented her with little to gain and much to lose. Nothing like the concern over hurricanes that had transformed Florida politics existed in Kansas; even to mention climate change meant being labeled an "Al Gore liberal." Those who thought about the issue at all were likely to accept the viewpoint articulated by right-wing leaders like Rush Limbaugh—that the alarm over climate change was nothing more than a ploy by liberals bent on controlling society.

Making the political prospects even worse for the governor was that the sponsor of the Sunflower project wasn't a large, out-of-state corporation but rather a homegrown Kansas cooperative with extensive political roots. As in most midwestern and southern states, rural electric cooperatives (RECs) like Sunflower are among the most well-organized and politically powerful institutions in state politics, enjoying connections to every legislative district. Many Kansans, especially the generation born in the early years of rural electrification, view their local REC with an affection that borders on reverence.

To understand the political power of the rural electric cooperatives in rural America, it's worth reflecting on farm life before electrification. To the big power companies that served urban areas, bringing power to the countryside didn't look like a profitable endeavor, so rural kids did their homework by kerosene lantern and farmers milked their cows by hand. The Edison Electric Institute, which represented investor-owned utilities, wrote, "only in the imagination ... does there exist any widespread demand for electricity on the farm or any general willingness to pay for it."

Left to their own devices, farmers started organizing rural electric cooperatives in the 1920s. After Franklin Roosevelt arrived in office in 1933, the movement took off. Looking for ways to jump-start an economy mired in depression, Roosevelt created the Rural Electrification Administration, which began lending money to RECs at low interest rates as well as giving them preferential access to cheap power from federal dams.

On paper, the original rural electric cooperatives looked very much like local food co-ops, and originally they elicited a comparable level of member involvement. In 1940 one observer wrote that co-op membership meetings "are not simply business sessions. They have an emotional overtone, a spiritual meaning to people who were so long denied the benefits of modern energy and convenience which had become a commonplace to their city neighbors."

Once the Rural Electrification Administration was organized, giving the RECs access to the deep pockets of the federal government, the movement lost much of its initial passion and became more institutionalized. Yet the fact that local boards continued to run cooperatives meant that a grassroots network remained available as a power base, affixed permanently to the Democratic Party. In 1964 the rural electrics came under attack as communist institutions, and federal funding for the movement became an issue in the presidential campaign between Lyndon Johnson and Barry Goldwater. Goldwater said that the Rural Electrification Administration had "outlived its usefulness." Johnson, on the other hand, had organized the REC that supplied electricity to his ranch in Texas. Not only was he a true believer in the benefits that the movement had brought to farm families, he also saw clearly that the thousand local co-op organizations operating in forty states were a politician's dream. Goldwater likewise noted the strength of this network. During a Senate hearing, he told a representative of the National Rural Electric Cooperative Association, "Within your organization you have a much more potent force at your fingertips than any source of pressure I have come in contact with since I have been here."

As time passed, what had once been among the most progressive organizations in America became in many ways one of the least. During the 1960s and 1970s, the local co-ops started moving into the big leagues, forming larger "generation and transmission" co-ops (G&Ts) to build billion-dollar coal plants and nuclear plants. In rural states like North Dakota and Kansas, managers of G&Ts like Basin Electric and Sunflower Electric became accustomed to the status and deference that Fortune 500 CEOs enjoy in larger states. Reacting to the red-baiting they had experienced in the 1960s, the co-ops seemed to lean over backward to emulate their private utility counterparts. With the advent of environmental legislation, they joined corporate America in lobbying against tougher strip mine regulations, clean air standards, and plant-siting legislation.

Farmers and ranchers facing destructive strip mining and intrusive power lines were dismayed when their efforts to organize landowner protection groups were vilified in REC publications and radio shows. The issue of climate change seemed to particularly enrage many co-op managers. Following the release of Al Gore's movie *An Inconvenient Truth*, the leadership of one co-op, Colorado's Intermountain Rural Electric Association (IREA), wrote to other utility managers asking for money to support the work of Patrick Michaels, a leading global warming denier. According to the letter, IREA had already given Michaels $100,000, and at least $50,000 more had been pledged.

Besides moving to the right on environmental issues, many

of the co-ops had drifted far away from basic cooperative principles. In a *Harvard Law Review* assessment of the problem, Congressman Jim Cooper of Tennessee wrote that the word "cooperative," with its implications of equal equity and member democracy, no longer fit the increasingly sclerotic reality. Cooper charged that the G&Ts that built and operated coal plants had become more intent on building economic empires for the benefit of their managers, boards, and circles of business cronies than on serving the needs of their members.

Sunflower Electric, which had been organized in 1957 by six western Kansas rural electric cooperatives looking for a reliable long-term power supply, exemplified the empire building syndrome. As proposed in early 2006, the three 700-megawatt coal plants scheduled for construction on the site of the existing 360-megawatt Holcomb plant would provide thirty times more power than needed by Sunflower itself—the lion's share would be sold to other co-ops, many located out of state. Tri-State G&T, based in Colorado, would actually own two of the plants. Most of the power from the third plant would go to Golden Spread Electric Cooperative of Texas.

At the outset, the handful of Kansas environmentalists who were aware of Sunflower's construction plans saw little chance of getting help from the Sebelius administration; indeed, there was no indication that the Sebelius administration would pursue any course other than to rubber-stamp the proposed plants. During her 2006 reelection campaign, the governor had ignored a Sierra Club request to support a moratorium on new coal-fired power plants, and the club had not endorsed her

As for any movement from the grassroots to oppose the project, early hearings on the Sunflower air permit were far from promising. At the first, held near the proposed construction site and packed with Sunflower employees, 90 percent of those testifying spoke in support of building the plant. So did local officials, who liked the project mainly because it would create two thousand construction jobs and 140 permanent jobs. At a second hearing, held in Topeka, supporters again outnumbered opponents.

At a third hearing, held in Lawrence, a college town in eastern Kansas, the tone began to shift. On this evening over three hundred people arrived to testify. As at previous hearings, Sunflower brought busloads of supporters to pack the room. Arriving early, the pro-company claque occupied the bulk of seats in the hearing chamber, forcing opponents who arrived later to stand in the halls.

Angered by the sense that they had been out-organized and that the process was rigged against them, opponents of the plant demanded a second night of hearings, and they quickly pulled together a coalition that included the Sierra Club, True Blue Women, the Sustainable Sanctuary Coalition, the Kansas Natural Resources Council, the University of Kansas Environs, and Concerned Citizens of Platte County. The coalition initiated a letter-writing campaign urging the governor to support wind power rather than coal plants.

As media attention steadily grew, so did attendance at each subsequent hearing. Hundreds of people showed up for Health Department hearings on the plant at the Kansas City Community College, and members of the coalition groups demonstrated at the governor's office in Topeka.

The involvement of environmentalists from eastern Kansas drew the ire of Sunflower executives. After the city of Lawrence took a position in opposition to the plant, Sunflower senior manager Steve Miller vowed, "I will personally make it my crusade to make sure all our western Kansas dollars are diverted as far away from Lawrence as they can be, because they have unfairly stuck their nose in western Kansas's business."

Inadvertently, Sunflower had changed the issue of the plant into an east-west controversy, a blunder in a state whose population tilts heavily toward its eastern counties. Although the company retracted Miller's comments, it was not able to do anything about the steady progression of the Sunflower issue onto the stage of national climate politics. State attorneys general from California, Connecticut, Delaware, Maine, New York, Rhode Island, Vermont, and Wisconsin all submitted objections to the permit application, complaining that the pollution from the plant would cancel out their own attempts at greenhouse gas reductions.

Even before Kansas made its decision, the Sierra Club had already filed a lawsuit challenging the state's refusal to hold an evidentiary hearing on the plants. Sierra followed the first lawsuit with a second in November, this time objecting that the Rural Utilities Service had failed to submit an environmental impact statement.

The tide was turning. In September 2007 Sunflower's opponents got some unexpected help when Colorado adopted a law requiring that rural electric cooperatives get 10 percent of their power from renewable resources. The Colorado legislation forced Sunflower to withdraw its application for the third unit of the project, which depended on commitments from Colorado utilities that were placed in question by the new legislation. In October the Sebelius administration announced its decision on the remaining two units. In a statement that made international headlines, Kansas Health and Environment commissioner Rod Bremby declared that the permit application for the remaining two units was being rejected on the basis of the U.S. Supreme Court's precedent in *Massachusetts v. EPA*. The era of blocking coal plants on explicit climate grounds had arrived.

Sunflower's response to the Bremby decision was swift and blunt: a broadside of attack ads, paid for by the newly formed pro-utility group Kansans for Affordable Energy, which sought to smear Governor Sebelius. Beneath photographs of Russian president Vladimir Putin, Venezuelan president Hugo Chávez, and Iranian president Mahmoud Ahmadinejad, the ads asked, rhetorically, "Why are these men smiling? Because the recent decision by the Sebelius Administration means Kansas will import more natural gas from countries like Russia, Venezuela and Iran."

Again, Sunflower's political instincts appear only to have made matters worse for the co-op. The effort to swiftboat the governor failed to intimidate Governor Sebelius, who called the attack campaign an affront to Kansan sensibilities. Sebelius released the following statement: "Anyone who would associate our state with the controversial and disreputable world leaders pictured in this ad fundamentally misunderstands and disrespects the people of Kansas. The ad is offensive to every Kansan, and the people of Kansas deserve an apology."

In the wake of the Bremby decision, the battle over Sunflower moved to the state legislature, where Sunflower allies, holding an easy majority, enacted legislation revoking Health and Environment commissioner Bremby's power to approve air permits. Sebelius vetoed the legislation, and the legislature tried the tactic again, both times failing to find the two-thirds majority needed to override the governor's veto. On May 29, 2008, the legislature ended its annual session without attempting to override a third veto by the governor.

Ironically, it was the election of Barack Obama that made it possible for one unit of the Sunflower plant to begin moving forward. In May 2009, immediately after Governor Sebelius left Kansas to take the position of Health and Human Services secretary in the Obama administration, the state's new governor, Mark Parkinson, reversed the Sebelius/Bremby position on the project, announcing that the co-op would be allowed to build a single 895-megawatt facility. Under the Parkinson plan, which was brokered with no public knowledge, most of the power from the plant would be exported to electric cooperatives in Colorado, while 200 megawatts would remain in Kansas. To offset pollution from the plant, Sunflower promised to shut several dirtier plants. The agreement also contained a provision sought by Sunflower and its allies to limit the Kansas Department of Health and Environment's power to regulate greenhouse gases and other pollutants. In exchange for allowing construction of Sunflower's coal plant, legislators were required to pass a bill enacting renewable energy measures sought by Parkinson.

For the coalition that had opposed the plant, it was a bitter loss, and Parkinson came under immediate fire. Critics charged that although the new governor touted the concessions made by Sunflower, the utility had already planned many of those concessions before striking the deal. Also as part of the agreement, Sunflower promised to decommission two outdated oil-fired power plants in Garden City, but according to the company, those oil burners had not actually been used in over two decades. Environmental groups charged that by stripping Kansas's top regulator of the discretion he had used to reject the Sunflower project in 2007, the deal would make it easier for other utilities to build coal plants in the state.

It remained an open question whether the Sunflower plant would actually be built. Like other proposed coal-fired power plants across the country, the project faced escalating construction costs, a tough climate for obtaining financing, and fading demand growth. Other sources of power, especially wind farms, looked competitive when compared with the cost of new coal plants. Meanwhile, legal teams at the Sierra Club and Earthjustice remained poised to seize any available handle to stop the project.

Despite what looked like defeat, at least in the short run, Sunflower opponents remained hopeful. Governor Sebelius was no longer in the state to lead the fight, but her vetoes had bought the anti-coal movement a critical window of time—perhaps just enough to win.

■

EIGHT

Direct Action

■

THE SEPTEMBER 2007 DECISION by Kansas governor Sebelius to block the Sunflower project galvanized the anti-coal movement, capping a summer of dramatic progress toward a moratorium on new coal plants. The onslaught of new coal plants was beginning to look less inevitable as project after project stalled or went off the rails. In Montana, Oklahoma, Kentucky, and Michigan, judges and regulators handed out rejection slips to coal plants. In North Dakota, Arizona, Washington, and New York, companies withdrew projects on their own initiative, citing such factors as rising costs, public opposition, and the prospect of carbon dioxide regulation. Citigroup downgraded the stocks of mining companies Peabody Energy, Arch Coal, and Foundation Coal Holdings, and that negative assessment further tarnished the prospects of companies seeking financing. In August U.S. Senate Majority Leader Harry Reid became the highest-ranking federal official to speak out against the building of coal-fired power plants. Prospects for stopping coal suddenly seemed much brighter than just a few months earlier.

On the other hand, governors like Sebelius and Crist and legislators like Reid were still the exception rather than the rule. The coal industry was well entrenched. Dozens of new coal plants remained in the pipeline, and existing plants were being run harder than ever. And governmental power notwithstanding, U.S. energy policy has always been driven primarily by a few dozen CEOs in the private sector. Given the existing power structure, many activists believed that they could never stop coal simply by participating in the prescribed channels of regulatory government. In their view, the regulatory structure itself tended to be more a way of giving environmentalists the illusion of involvement while rubber-stamping coal projects.

But if the regulatory system was a distraction at best and a fraud at worst, what strategy could activists follow? One answer was "direct action," the sort of peaceful but confrontational tactics used most famously during the civil rights movement. Opponents of strip mines and high voltage transmission lines had used confrontational tactics before—sometimes peacefully, other times not. In 1965 Ollie Combs, a 61-year-old widow, sat down in front of a bulldozer along with her two sons to stop their Honey Gap, Kentucky, home from being mined by the Caperton Coal Company. A newspaper photograph showing Combs eating Thanksgiving dinner behind bars produced a public outcry and led to the founding of Appalachian Group to Save the Land and People, which by 1972 was staging organized non-violent civil disobedience actions, including a January 1972 strip mine occupation by 20 women in Knott County, Kentucky, that received national media attention.

Not all resistance efforts were nonviolent. In Knott County, a diesel-powered shovel owned by Kentucky River Coal was dynamited in April 1967, and another large shovel was blown up two months later at a Kentucky Oak operation nearby. In Perry County, Kentucky, saboteurs dynamited a grader belonging to the Tarr Heel Coal Company and snipers exchanged gunfire with workers. Later that summer, carbon nitrate was used to destroy two trucks, an auger, and a bulldozer at the same location. Local strip mine opponents in Knott County also formed a "conservation group" called the Mountaintop Gun Club, which assisted landowners in setting up shooting ranges to dissuade miners from encroaching on their property.

In Pope County, Minnesota, a years-long confrontation over a large powerline extending from the North Dakota coal fields to Minnesota's urban centers erupted into a full-scale rebellion during 1978 between hundreds of farmers armed with tractors, manure spreaders, and ammonia sprayers and two hundred state troopers attempting to protect surveying operations for the line. After the powerline was built, "bolt weevils" toppled numerous towers and bullets from high-powered rifles damaged insulators and transmission cables. Despite intensive deployment of utility security personnel and police resources, including high-speed helicopters, no arrests were ever made.

During the current wave of opposition to coal, protesters have steered firmly toward nonviolent tactics. In 2003 a group of protesters called the Rocky Top affinity group, affiliated with Katúah Earth First, locked themselves into concrete-filled steel barrels, blocking the entrance to the Zeb Mountain mine in Tennessee. The three protesters, "john johnson," Dan Anderson, and Matthew Hamilton, were arrested and released that day. Near the mine, another group climbed a 150-foot billboard off Interstate 75 and hung a banner reading "Stop Mountaintop Removal." By 2007 the pace of direct action protests was quickening. At least seventeen such protests took place that year, rising to forty-two in 2008, and thirty-five in the first half of 2009. Appendix A provides brief descriptions of about a hundred direct action protests.

Direct action has a long history, and a considerable amount of thinking has been devoted to understanding how its tactics—particularly actions that deliberately defy the law—can have such a salutary effect, especially when other, seemingly more reasonable measures have failed.

It's not at all obvious that this should be true. As anyone who has been to a direct action protest can attest, a lot of what goes on involves clumps of bored police standing around, a few protesters sitting on the ground connected by odd-looking PVC pipes and other devices, some signs and chanting, and a smattering of onlookers and press. Sometimes passersby are supportive, sometimes not.

The idea that this sort of thing can actually produce far-reaching political success may seem counterintuitive, but a paper published in June 2007 by University of Washington sociologist Jon Agnone argues persuasively that it actually does. According to the paper, "Amplifying Public Opinion: The Policy Impact of the U.S. Environmental Movement," which summarized data on trends in public opinion, the incidence of protest actions, and the passage of environmental legislation, the evidence shows protest to be more effective in spurring legislators than either public opinion or institutional initiatives. Based on a national survey of protests kept during the period 1960 to 1998 by the *New York Times*, the occurrence of protests increased the passage of environmental legislation by 9.5 percent. Public opinion by itself also influences legislation, but protest "raises the salience of public opinion for legislators," according to Agnone. He describes the effect as "amplification."

Few anti-coal activists who participate in direct action protest have read Agnone's research. Their motivation is based more on a sense of moral conviction that the urgency of the climate crisis and the other effects of continued coal use compel them to do more than just write a letter or sign a petition. Among the many groups that sponsor direct action protests, perhaps the largest is Rising Tide, which began in Europe before jumping the pond to the United States.

To learn more about the group, which has no paid staff and no central office, I signed up to attend the West Coast Climate Convergence in Skamakowe, Washington, and in early August 2007 I found myself driving along the north shore of the Columbia River, a lushly forested area where greenery flooded out from both sides of the road like tendrils in a rainforest. Blackberry bushes loaded with ripe fruit tumbled out onto the asphalt. The event was held on the Wahkiakum County fairgrounds in the southwestern corner of the state. When I arrived, I found a bulletin board strewn with sheets of announcements and a sign-in table set up under a canvas tarp. Nearby was an open-air kitchen where half a dozen members of the Seeds of Peace cooking collective chopped vegetables and stirred industrial-sized pots of food. Otherwise, there seemed to be few people in attendance and no visible headquarters or apparent leadership structure. I carried my gear to a grassy encampment area, set up my tent among several dozen other tents of assorted designs, stashed my sleeping bag and backpack in the tent, and went back to the bulletin board to see if I could find a schedule of events.

It took me awhile to adjust to the sensation that the convergence lacked any center or sense of focus. Eventually, I figured out that the nervous system of the camp revolved around short, productive meetings held each morning on the steps of the fairground offices about fifteen minutes before breakfast. At the meetings, a handful of organizers worked their way through the business of the day, speaking a patois loaded with jargon like "bottom lining" and "consensing." There was little wasted motion; no particular person seemed to be in charge.

It was the famed leaderless coordinating style of the youth climate movement. Although direct action is most often associated with protesting *against* something, the youth climate movement can also be seen as a large, far-flung experiment in new ways to run groups and make decisions without top-down hierarchies and arbitrary authority. This puts the movement in the wide tradition of anarchist, anti-authoritarian social innovation. Interestingly, many concepts from that tradition, such as "open space" meeting theory, seem to quickly hop the fence into the corporate world.

Most of the activity of the Climate Convergence in Skamakowe took place in self-organized sessions held in buildings and outside under the trees. Hundreds of people shared information on topics as varied as organic gardening, mountaintop removal mining, multiracial organizing principles, "tall bike" mechanics, technical tree-climbing skills for direct action, and even "insurgent rebel clown army training." A particularly lively workshop was "radical cheerleading chants," taught by Canadian organizer Mike Hudema. Workshops began early and continued into the evening. Late evenings were spent in lectures or films. Later still began the informal strategy sessions and guitar playing around scattered bonfires.

During a rare lull, I sat at a picnic table quizzing a young Englishwoman named Sophie about the origins and history of Rising Tide. With a ready grin and the slightly preoccupied air of a bookworm, Sophie seemed more like a college student taking a short study break than a seasoned organizer. In fact, she had spent the better part of the previous three years traveling the breadth of the Anglophone world to organize various sorts of direct action protests, steadily building the Rising Tide network. In County Mayo, Ireland, she supported the Rossport Five, a group of protesters jailed for three months for resisting a liquefied natural gas pipeline. In 2005 at Gleneagles, Scotland, she participated in a camp of 5,000 people protesting the G8 summit. That experience gave birth to the notion of climate camps, the first of which took place the following year along with a large blockade at the Drax Power Station in North Yorkshire, England. Sophie missed Drax, instead participating in a blockade at a coal mine slated for an alpine wetland in Happy Valley, New Zealand. Most recently, Sophie had been at Black Mesa, Arizona, working in support of Hopi and Navajo elders who were being relocated from their land.

Sophie explained that Rising Tide had arisen out of the frustration of climate activists with the bureaucratic, sluggish, and corporate-dominated Kyoto Protocol process. At The Hague in 2000, protesters invaded the Kyoto conference, denounced the proceedings as a trade fair for industry, and threw a berry pie into the face of the chief U.S. negotiator. Surprisingly, Michael Zammit Cutajar, executive secretary of the conference, took the action in stride. Rather than order that the demonstrators be arrested, he applauded them, saying, "I hope the impatience if not the methods of the protesters will get transmitted to the negotiators."

Not everyone approved of Rising Tide's confrontational tactics. To many activists associated with mainstream environmental groups, actions like sit-ins and banner hangs that cross the line into civil disobedience served simply to alienate the general public and potential corporate allies. Better to strike a business-friendly, "reasonable" tone. But Cutajar's appreciation of the action at The Hague was not untypical, even among participants in the "inside" game of climate negotiations. To them, an "inside/outside" combination of tactics could be useful both in climate negotiations and in legislative arenas.

Two years after Rising Tide's disruption of negotiations at The Hague, members of the movement met in Barcelona, where they hammered out a comprehensive statement of principles that put issues of social and global equity at the core of solutions to climate change. The statement also endorsed direct action tactics as the key tool for challenging corporate opponents, and it committed the Rising Tide movement to a nonhierarchical structure.

If the climate war is someday judged to be won, I wonder how much of the credit will go to the organizers who, like Sophie, have scrambled to knit activists from around the world into a coherent movement. Months later, as I watched Rudolph Giuliani and Sarah Palin mock Barack Obama's community organizing experience, I reflected further on the organizers I had met in the climate movement. "I guess a small-town mayor is sorta like a community organizer," Palin told the delegates at the Republican convention, "except that you have actual responsibilities."

Giuliani won laughs when he followed up, sarcastically: "He 'worked' as a community organizer!"

Actually, Giuliani's own run for the presidency might have gone better if he had adopted some of the traits and methods Obama had picked up during his community-organizing days. While the Giuliani campaign fell prey to clashing egos and staff infighting, Obama filled his campaign staff with meticulous, hardworking, no-drama personalities.

In the climate realm the ingredients for success were no different. I was struck by the characteristics that successful organizers seemed to share: energy, humor, nerve, and lack of pretension. The cultural stereotype of the shrill, ineffectual radical could not be further from the mark. Sophie was definitely serious about her work. "We have very little time to turn the global climate crisis around," she said grimly.

Yet what might have otherwise been an off-putting sort of intensity was leavened by a mischievous joie de vivre. After supper, while some attendees at the convergence finished their curried cauliflower soup and others lined up for a helping of apple crisp, Sophie stood up to make an announcement, "Urgent matter. Attention! I have to leave tomorrow early, so tonight is my last chance. I've got an awesome mix on my iPod and tonight I intend to dance my ass off in the 4H hall. Ten p.m.!"

As I crawled into my sleeping bag that night, the pulsing beat of Sophie's music rocked the Skamakowe fairgrounds, and I marveled at the energy of youth.

Not long after returning from the Rising Tide conclave, I found myself in an urban forest, surrounded by massive skyscrapers in downtown San Francisco. As part of a nationwide action organized by Rainforest Action Network, I had been sent out along with another activist to stage mini-actions in the North Beach neighborhood. Across the city, other teams had been assigned other neighborhoods.

Our target was ATMs and branch bank offices of Bank of America and Citibank. At each location, we would block off an ATM machine or the doorway to a bank office with yellow tape, upon which the words "Climate Crime Scene" had been printed. We'd admire our handiwork, take a few snapshots, then move on to the next location.

As street theater, the action embodied a certain humor. It certainly caused no harm to the banks' business, yet the fact that scores of branch offices around the United States were targeted on the same day was guaranteed to get the attention of the highest bank executives. The goal was to persuade Bank of America and Citibank to stop financing coal plants and mountaintop removal mines and shift their lending toward clean energy. In his book *Coming Clean*, which recounts Rainforest Action Network's successful campaign against Citibank's financing of rainforest logging, RAN's executive director Mike Brune describes why even mild protest aimed at undermining a corporation's public image often gets results:

> High school and college students are red meat for banks. Once a bank starts doing business with a young person, it won't let go. Banks aim to entice students with their first credit card, and then as time passes, ply them with student loans, auto loans, mortgages, investments accounts, retirement plans and so on. Each semester at high schools, universities, and college campuses across the country, Citi employees would arrive on campus to sign up students as new credit card customers. It was a golden opportunity for our campaign work.... As the Citi campaign continued, we placed an advertisement in the *New York Times*. The headline "Did you know someone is using your credit card without your authorization?" ran above pictures of clearcut forests, oil pipelines, and pollution-belching smokestacks. ... We began to receive thousands of cut-up cards by mail.

But what if RAN succeeded and convinced the two banks to stop underwriting coal projects? Couldn't such projects simply seek funding somewhere else? Perhaps, but by squeezing financial channels, project costs would rise, forcing the economics of energy to shift toward greener sources.

At least that was the idea. It also occurred to me that another motive for picking banks was the simple reality that they presented a more convenient target for urban activists than remote plants and mines. Of course, on the spectrum of direct action, creating a whimsical piece of street theater at an ATM in San Francisco is missing one key ingredient of the most effective direct action: personal risk. In Appalachia, where protesters faced violent retaliation by police or coal workers, or the threat of heavy-handed prosecution, such risks were very real. In Knoxville, Tennessee, police had used choke holds and pain compliance when forty-five Mountain Justice activists, some clad humorously in animal costumes and playing marching band instruments, descended on a shareholders meeting of the National Coal Corporation. At another demonstration at a National Coal mine site, company workers had threatened protesters and attempted to ram them with a car. In North Carolina, protesters at Dominion's Cliffside Plant were tasered and placed in pain compliance holds. In Ohio, police peppersprayed protesters conducting a sit-in at the headquarters of American Municipal Power. In West Virginia, mine workers threatened and assaulted anti-coal activists; houses of activists were been shot at, vandalized, and even fire-bombed.

On September 15, 2008, in Wise County, Virginia twenty protesters entered the construction site of a Dominion Resources coal-fired power plant and locked their bodies to eight large steel drums, two of which had operational solar panels affixed to the top that illuminated a banner reading "Renewable jobs to renew Appalachia." Outside the construction site, others sang and displayed a large banner with the message "We Demand a Clean Energy Future." Eleven were arrested and charged with misdemeanors. But two of the arrestees, Hannah Morgan and Kate Rooth, were charged with ten more crimes than the other defendants, including "encouraging or soliciting" others to participate in the action and "obstruction of justice." If convicted, the two faced up to fourteen years in prison. Confronted with that prospect, they agreed to a plea bargain.

Climate scientist James Hansen had offered to testify on behalf of Rooth and Morgan if the case had gone to trial. He had earlier testified at a trial for the "Kingsnorth 6," a group of Greenpeace members who occupied the 200-meter smokestack of the Kingsnorth Power Station in the United Kingdom. Using a "lawful excuse" defense—an argument that the crimes they had committed were intended to prevent a greater wrong—the Kingsnorth 6 had won aquittal. About the Rooth and Morgan case, Hansen wrote:

> If this case had gone to trial I would have requested permission to testify on behalf of these young people, who, for the sake of nature and humanity, had the courage to stand up against powerful "authority." In fact, these young people speak with greater authority and understanding of the consequences of continued coal mining, not only for the local environment, but for the well-being of nature itself, of creation, of the planet inherited from prior generations.

> The science of climate change has become clear in recent years: if coal emissions to the atmosphere are not halted, we will drive to extinction a large fraction of the species on the planet. Already almost half of summer sea ice in the Arctic has been lost, coral reefs are under great stress, mountain glaciers are melting world-wide with consequences for fresh water supplies of hundreds of millions of people within the next several decades, and climate extremes including greater floods, more intense heat waves and forest fires, and stronger storms have all been documented.

> Our parents did not realize the long-term effects of fossil fuel use. We no longer have that excuse. Let us hope that the courage of these young people will help spark public education about the climate and environmental issues, and help us preserve nature for the sake of our children and grandchildren.

■

The Education of Warren Buffett

■

RALLIES, LAWSUITS, PETITION DRIVES, sit-ins, street theater—
these are not just abstract tactics. They are aimed at changing
decisions in the real world. But do they work? As described in
the previous chapter, sociologist Jon Agnone has come up with
compelling data to support his "amplification model of public
policy impact," which explains how citizen protests end up
affecting legislation. That would be more encouraging for the
anti-coal movement if the key decisions about energy policy in
the United States were actually determined by the democratic
process. But many—perhaps most—are not. Rather, they are
made in executive suites and boardrooms by decision makers
who never run for public office. These men and women are
legally accountable to maximize their corporations' bottom
lines, not to any broader concern. Their power is buttressed
by a judicial and political framework systematically rigged to
protect their prerogatives. Summarizing this reality, former
UN ambassador Andrew Young once said, "Nothing is illegal
if 100 businessmen decide to do it."

Yet some activists saw that very concentration of power in

TABLE 1: KEY PRIVATE SECTOR DECISION MAKERS ON COAL

CEO	COMPANY/ENTITY	UNITS*	CAPACITY (MW)†
Michael G. Morris	AEP	63	27,636
David M. Ratcliffe	Southern Company	68	26,610
James Rogers	Duke Energy	70	18,578
Tom D. Kilgore	TVA	63	17,647
Gary L. Rainwater	Ameren	31	10,719
G. Abel / W. Buffett	MidAmerican (Berkshire)	29	10,281
John W. Rowe	Exelon	21	9,415
Richard C. Kelly	Xcel Energy	30	9,021
David W. Crane	NRG Energy	26	8,657
Anthony J. Alexander	FirstEnergy	36	8,495
Thomas F. Farrell II	Dominion	32	8,417
Wulf H. Bernotat	E.ON	29	8,347
Mark M. Jacobs	Reliant Energy	26	8,133
Anthony F. Earley Jr.	DTE Energy	22	7,997
William D. Johnson	Progress Energy	23	7,924
Paul J. Evanson	Allegheny Energy	22	7,636
David Campbell	Luminant	9	6,137
James H. Miller	PPL	13	5,981
Paul Hanrahan	AES	29	5,406
Edward R. Muller	Mirant	18	4,075
William D. Harvey	Alliant Energy	30	4,055
J. Wayne Leonard	Entergy	5	4,014
Bruce A. Williamson	Dynegy	12	3,755
Paul M. Barbas	DPL	11	3,521
Robert C. Skaggs Jr.	NiSource	10	3,470

*Units: Number of coal-fired generating units in the United States in 2005.
†Capacity: Coal-fired generating capacity in 2005 expressed in megawatts (MW).

individual hands as a potential point of leverage. As shown in table 1, a mere two dozen chief executive officers control the fate of over 70 percent of all coal-fired power generation in the United States. Might it be possible that subjecting these men to direct pressure—including reminding them that destroying the climate would affect their own children and grandchildren—could produce some sort of awakening? Alternatively, even if those with inordinate private power were impervious to the fate of the planet, might they at least care about their public images?

James Hansen, for one, seemed to think that coal and utility executives could be swayed by direct appeals. His message was blunt: "CEOs of fossil energy companies know what they are doing and are aware of long-term consequences of continued business as usual. In my opinion, these CEOs should be tried for high crimes against humanity and nature."

Hansen also sought to engage fossil executives directly. In a letter to Duke Energy CEO James Rogers, he wrote "a plea for cooperation and leadership":

MARCH 25, 2008

To: Mr. James E. Rogers, Chairman, President and Chief Executive Officer, Duke Energy
FROM: Jim Hansen, Columbia University Earth Institute
SUBJECT: A Plea for Cooperation and Leadership

Mr. Rogers, as a leader in the electric power industry, your decisions will affect not only energy bills faced by your customers, but the future planet that your children and grandchildren inherit. If you insist that new coal plants are essential for near-term power needs, you may submit your company and your customers to grave financial risk, and leave a legacy that you will regret.

Scientific evidence of human-made climate change has crystallized, and it has become clear that continued emissions carry great danger. These facts fundamentally change liabilities. And liabilities will be increased by any "success" of industry efforts to confuse the public about the reality and likely consequences of human-caused climate change and to promote false "solutions" such as new "cleaner" coal plants.

Surely the number of people pressing these legal cases will grow, and they will be inexorable in pursuing justice. And assuredly, in the long run, the energy companies will lose the legal battles.

Unfortunately, although the public will ultimately hold polluters accountable, it will not necessarily be soon enough or have enough impact to prevent environmental and human disasters. It may drag out as in the tobacco case, but with much more serious consequences.

Mr. Rogers, this is a path that, for the sake of our children and grandchildren, we cannot follow. Enlightened leadership is desperately needed in planning our energy future. As a captain of industry, you can help inspire this country and the world to take the bold actions that are essential if we are to retain a hospitable climate and a prosperous future. I am reaching out to you, Mr. Rogers, because you are uniquely positioned to influence others in your industry, and because your statements suggest that you comprehend the gravity of the problems we face.

Hansen's words had no discernable effect on Rogers, who continued pursuing two coal plants, one at Cliffside in North Carolina and the other at Edwardsport in Indiana.

Other activists chose a more blunt tool: ridicule. Wearing black suits and top hats that looked like smokestacks while ostentatiously sipping martinis, Billionaires for Coal held mock conventions in New York City and Houston celebrating the coal investments of Merrill Lynch. In Richmond, Virginia, they partied in front of Dominion Power's headquarters; the We Love Money String Band provided entertainment and assured the audience that "we're only in it for the money."

Beginning in 2004 Rising Tide and the Energy Action Coalition organized Fossil Fools Day each April, staging humorous actions and bestowing mock awards known as Foolies on energy executives. One coal mogul who managed to escape public ridicule was Warren Buffett. His house in Omaha hadn't been picketed; he hadn't received a Foolie; and no one had shoved a key lime pie in his face (a common tactic in England). Yet only a handful of men oversaw more coal plants than MidAmerican Energy, a subsidiary of Warren Buffett's holding company, Berkshire Hathaway. In all, MidAmerican Energy's operations included twenty-nine coal plants, and the company was planning at least seven more. One of those was a 760-megawatt unit in Iowa, which in the summer of 2007 was about to go online. A second was planned for Delta, Utah, and a third for Rock Springs, Wyoming. At least four others would be built in the Rocky Mountain region at locations that had not yet been determined.

Given Buffett's general experience of being worshipped by grateful stockholders, I wondered how he might react to the kind of derisive attention that groups like Billionaires for Coal were so good at dishing out. Before I had a chance to find out, however, Buffett had quietly exited the stage, canceling all the coal plants (with the exception of the now-completed Iowa plant) that he had been planning to build just a year earlier. In late 2007 his PacifiCorp subsidiary told regulators it planned to supply future electricity demand growth through geothermal, wind, solar thermal, compressed air storage, conservation programs, and natural gas. Plans to build new coal plants were off the table.

What accounted for Buffett's change of direction? We can only venture some educated guesses, since the investment guru declined to reflect publicly on the decision, a notable departure from his usual practice of explaining his major moves in an annual letter to Berkshire Hathaway's shareholders. This was a big disappointment. Buffett is famous for the candor and clarity of his commentaries, and his insights into American social and business trends can often be profound. Due to his legendary record as an investor and corporate strategist, his discourses on business are often studied like tutorials across the business world. When Buffett exited coal without explanation, an opportunity for helping the world of commerce begin to conceive of a post-carbon future was unfortunately missed. Reconstructing the sequence of events that led Berkshire Hathaway and MidAmerican Energy to abandon their plans for new coal plants may make it possible to arrive at the underlying rationales for the decisions, thereby articulating the emerging business case for moving beyond coal.

Before canceling the plants he had intended to build, Buffett seemed to love coal. His involvement with building coal plants began when Berkshire Hathaway bought MidAmerican Energy Holdings in 1999. MidAmerican was a big operator of coal plants, and as natural gas prices edged toward a huge leap upward—bringing coal back into favor—the purchase of MidAmerican appeared to be a typically savvy Buffett move.

In 2006 Buffett picked up another utility, PacifiCorp, which included Rocky Mountain Power and operated in California, Idaho, Oregon, Utah, Washington, and Wyoming. Again, it seemed like a smart play, bringing MidAmerican's expertise with building and running coal plants to a region of the country with lots of coal. Sure enough, in the fall of 2006, PacifiCorp presented regulators with plans for half a dozen coal plants to be built in Utah and Wyoming over the coming twelve-year time period, representing approximately 3,000 megawatts of new capacity.

The first sign that a major change was afoot in Buffett's coal strategy came in May 2007, when PacifiCorp released a new iteration of its Integrated Resource Plan, a massive document periodically provided to utility regulators in Oregon. Buried in the document was a huge change in PacifiCorp's coal strategy: four coal plants that had been shown in previous versions of the plan were now omitted.

It is clear that the cancellation of these first four plants was not the result of any sort of personal awakening on Buffett's part about the urgency of climate change. Rather, PacifiCorp was bowing to pressure by state governments in California, Oregon, and Washington. In a footnote to the May Integrated Resource Plan, company planners listed the following factors: "the Oregon PUC rejection of the 2012 RFP [Request for Proposals] for baseload resources and issuance of new IRP guidelines (January 2006), adoption of renewable portfolio standards in Washington, California's adoption of a greenhouse gas performance standard, and introduction of climate change legislation in both Oregon and Washington."

Further evidence that the elimination of four coal plants from PacifiCorp's plan was driven by outside pressure from state regulators rather than by a change in the sentiments of Buffett or his executive staff can be gleaned from comments by Charles Munger, vice chairman of Berkshire Hathaway, at the company's annual meeting in May 2007. In response to a question about global warming, Munger said, "What we are really talking about with global warming is dislocation. Dislocations could cause agony. The sea level rising would be resolved with enough time and enough capital. I don't think it's an utter calamity for mankind, though. You'd have to be a pot-smoking journalism student to think that."

Like Munger, Buffett's partner-in-philanthropy Bill Gates, Jr., also downplayed the urgency of the climate crisis. The views of Gates on social concerns are relevant to gleaning Buffett's views; after all, only months earlier Buffett had announced that he was donating approximately 70 percent of his fortune to the Bill and Melinda Gates Foundation. In April 2007 Gates told a forum in Beijing, "Well, fortunately climate change, although it's a huge challenge, it's a challenge that happens over a long period of time. And so [according to] most of the forecasts about by the year 2100 the ocean will have risen perhaps a foot and a half. You know, we have time to work on that."

It is hard to imagine both Munger and Gates displaying such complacency about climate change if Buffett felt much differently, at least as of the spring of 2007. But by the end of the year, Buffett would cancel two more plants for reasons that seem less clearly driven by pressure coming from state regulators. The two cancellations were announced on November 28, 2007, in a letter sent by PacifiCorp to regulators in Utah and Oregon. The explanation was terse: "Within the last few months, most of the planned coal plants in the United States have been canceled, denied permits, or been involved in protracted litigation."

The reference to litigation suggests that the management of MidAmerican had been watching the ongoing fights over coal plants in states like Kansas, Minnesota, Delaware, Texas, and Florida and had concluded that it preferred not to enter that sort of legal gauntlet. In fact, one of the two proposals, the Intermountain Power Project Unit 3, had already landed in court. The majority cosponsors of Intermountain Units 1 and 2 were a group of six California cities: Los Angeles, Pasadena, Anaheim, Burbank, Glendale, and Riverside. Prohibited by California climate laws from using Unit 3's power, the six cities had decided to actively block the new unit. MidAmerica's PacifiCorp unit had originally threatened its municipal partners in the project with legal action to force their participation. But with its November announcement, PacifiCorp waved the white flag.

The second of the two final cancellations, a new unit at the existing Jim Bridger station in Wyoming, was not the subject of any litigation. Nevertheless, Buffett's managers may have informed him that even in coal-friendly Wyoming, the project would inevitably become controversial. With every passing month, the breadth of the anti-coal movement in the Mountain States was growing. At the radical end of the spectrum, Cascadia Rising Tide, Stumptown Earth First!, and the Convergence for Climate Action had already blocked PacifiCorp's headquarters in August 2007 with a "human dam." As youth-based direct action continued to ramp up, such protests were certain to become more and more frequent.

Pursuing more conventional political channels, local citizens in Utah with the group Sevier Citizens for Clean Air and Power were beginning to push for a grassroots initiative that would mandate a public vote on any new coal-fired power plant in the area. Grassroots activity like that in Sevier could be found in every state in the Mountain States region, much of it directed at PacifiCorp. Meanwhile, mainstream environmental and civic groups were investing hundreds of staff and member hours in state utility oversight proceedings, especially in Oregon. Among the most active of such groups were the Northwest Energy Coalition, Citizens' Utility Board of Oregon, Ecumenical Ministries of Oregon, the Renewable Northwest Project, Western Resource Advocates, and Sierra Club Utah Chapter. The Northwest Energy Coalition alone represented over one hundred organizations, including solar companies, public power agencies, environmental groups, civic groups, and housing authorities.

Mayors in several Rocky Mountain states were also speaking out against new coal plants, including the mayor of Park City, Utah, Dana Wilson, who wrote a letter to Buffett expressing the city's opposition to new coal plants.

Even some members of the business community were beginning to apply public pressure on Buffett to drop his coal plans. In Salt Lake City, commercial real estate broker Alexander Lofft initiated a petition drive that collected 1,600 signatures from a "collection of citizens, business owners and managers, service professionals, public servants, and organization representatives ... your friends and new customers here in Utah." In a letter accompanying the petition, Lofft's ad hoc group wrote that any further expansion of coal generation in Utah would "compromise our health, obscure our viewsheds, shrink and contaminate our watersheds, and thin out our most beloved snowpack." It continued: "Our attractiveness as a place to live and work is also threatened, and so is our economic competitiveness as a major metro area and a state, compromising our recent gains in income and property values."

It is hard to avoid the conclusion that the accumulation of pressure from so many points on the political spectrum did indeed have a telling effect on Buffett as well as on his lieutenants, MidAmerican chairman David Sokol and PacifiCorp chairman Gregory Abel. According to Buffett's 2008 annual letter to his shareholders, decisions on "major moves" at MidAmerican are made only when he, Sokol, and Abel "are unanimous in thinking them wise." Sokol, like Buffett an Omaha native and resident, is a seasoned utility executive whose résumé included building geothermal facilities (CalEnergy) and coal (the controversial

Council Bluffs Unit 4 power plant). In the midst of the budding controversy over Buffett's coal plans, Sokol announced that he was stepping down as CEO of MidAmerican in order to assume a larger role in Berkshire Hathaway. Industry observers speculated that Sokol was being groomed to eventually replace Buffett himself, either as the company's leader or as one of a troika of leaders.

As he prepared to step into the shoes of a man who is widely seen as a de facto statesman of American business, it is conceivable that Sokol preferred not to invite the image problems that a protracted fight over coal might entail.

On the whole, Buffett's change of direction on coal shows a striking similarily to his change of direction in the early 1990s from a pro-tobacco investment posture to a policy that was much more wary of such investments. In 1987 Buffett told John Gutfreund of Salomon, "I'll tell you why I like the cigarette business. It costs a penny to make. Sell it for a dollar. It's addictive. And there's fantastic brand loyalty."

By 1994 Buffett's statements on tobacco had shifted notably. He told Berkshire Hathaway's annual meeting that tobacco investments are "fraught with questions that relate to societal attitudes and those of the present administration... I would not like to have a significant percentage of my net worth invested in tobacco businesses."

What we can conclude from all this is that stigmatizing coal and putting direct pressure on utility executives probably does work, especially when it comes on top of other forms of pressure. Just as he had with tobacco, Buffett got the message that America was about to start sending coal-boosting executives to the woodshed. For other CEOs, either the "Aha!" moment took somewhat longer to sink in or the inertia of their pending investment programs was too powerful to quickly overcome. Less quick on their feet, Dynegy's Bruce Williamson, Duke's James Rogers, and TVA's Tom Kilgore would all be soon receiving their Fossil Fool, Corporate Scrooge, and other badges of dishonor.

What caused Warren Buffett to cancel six coal plants? All of the following must be given their due: (1) strict carbon dioxide emissions standards enacted in California and Washington; (2) renewable portfolio standards in California and Washington; (3) climate change legislation in Oregon, California, and Washington; (4) rising construction costs for coal plants; (5) increased competitiveness of alternatives such as wind; (6) the prospect of national carbon legislation; (7) Oregon's integrated resource planning process; (8) regulatory participation by mainstream environmental groups such as the Northwest Energy Coalition; (9) litigation and the threat of litigation by groups such as the Sierra Club; (10) a medley of citizen actions that "raised the negatives" for coal, including anti-coal statements by mayors in several Rocky Mountain cities, direct action protests by groups such as Rising Tide, Alexander Lofft's petition drive in Utah, personal advocacy by prominent figures such as James Hansen, and concerted campaigns to place a public stigma on coal, such as the Foolie awards.

Although activists will always carry on debates among themselves about which tactics are the most effective, the real lesson to be gleaned here is the value of the widest possible range of approaches and the involvement of multiple organizations and sectors of society. This is the "swarm" in operation—the best hope for winning the war to stop coal and prevent needless climate chaos.

■

TEN

Progress Report: 59 Coal Plants Down

■

BUFFETT'S RADAR WAS ACCURATE: by the late fall of 2007 a groundswell of opposition to coal was undeniably emerging across the country. Mainstream political bodies such as city councils, state legislatures, municipal utility districts, and Alaskan Native corporations were taking positions in favor of a moratorium on new coal plants. In Salt Lake City, Mayor Rusty Anderson expressed vocal opposition to new coal plants, as did mayors in Pocatello, Idaho; Park City, Utah; and elsewhere. In Charlottesville, North Carolina, Mayor David Norris posted aerial photographs of mountaintop removal mining on his blog, and the city council passed a resolution urging the Commonwealth of Virginia to institute a ban on new or expanded coal plants. In Montana, local politicians found out just how unpopular coal plants had become after Southern Montana Electric Generating and Transmission Cooperative went looking for cities to share the power from its proposed Highwood Generating Station. The city council of Helena voted not to purchase power from the plant, citing emissions concerns and other factors. Missoula mayor John Engen had previously won city council approval to purchase electricity from Highwood, but after receiving hundreds of e-mails and phone calls from angry constituents, he reversed his position.

"Coal is a double-edged sword," Mayor Engen told the *Washington Post*. "I sort of felt both edges."

Spontaneous uprisings against proposed coal plants were becoming remarkable not only in their frequency but in their creativity. In Wiscasset, Maine, a coalition of local environmentalists and lobster fishers organized a flotilla of thirty boats to demonstrate against a proposed coal gasification plant that would have required a constant traffic of coal barges on the Sheepscot River, disrupting lobstering and fishing operations. Ignoring promises of hundreds of new jobs, area residents rejected the plant in a local referendum by a vote of 868–707. Although the Connecticut-based real estate developer who was seeking to build the Wiscasset plant vowed to press on, that possibility became increasingly remote after opponents secured state legislation instituting a three-year statewide moratorium on coal gasification and a carbon emissions standard thereafter that could only be met by plants that captured and stored their carbon dioxide. Since Maine lacked any geological formations suitable for carbon sequestration, the likelihood of further coal plants in the state dwindled to insignificance.

On the other side of the country, a consortium of twenty public power organizations known as Energy Northwest had sought to join with a group of private utilities to build a coal plant in the Port of Kalama, Washington. The group applied for permits in August 2006, and throughout 2007 a wide array of groups led by the Northwest Energy Coalition worked to oppose the project. The coalition had already gained leverage by securing passage of a strict standard on greenhouse gas emissions. Now, the fight shifted to whether the Kalama Plant, which claimed it intended to sequester its carbon underground, would actually follow through on the promise. In November 2007 the state Energy Facility Site Evaluation Council voted unanimously to reject the Kalama project, declaring that the sponsor's statements about carbon sequestration amounted to "a plan to make a plan."

Once confined mainly to rural areas, protests against coal were becoming an urban phenomenon as well. One of the most outrageous examples of urban pollution from coal was in Chicago, where two plants, Fisk and Crawford, both owned by Midwest Generation, spewed emissions into a densely populated Latino barrio known as Little Village, home to ninety-five thousand people. Because of their age, the 40-year-old Fisk plant and the 50-year-old Crawford plants were exempt from federal pollution regulations. A 2001 study by the Harvard School of Public Health found that the two plants were causing forty premature deaths each year, as well as 2,800 asthma attacks. Samuel Villaseñor and other activists at the Little Village Environmental Justice Organization (LVEJO) came up with the idea of using Chicago's bid to host the 2016 Olympics as a point of leverage for closing down the two plants. The Coalympics, held across the street from the Fisk plant, included the "Coal Power Plant Hurdle." Participants in the event competed by jumping over miniature coal power plants rather than regular hurdles.

"If our mayor isn't willing to represent the true people of Chicago, then we will represent ourselves with the International Olympic Committee and let them know that it's not as pretty a picture as he paints it to be," said Kimberly Wasserman Nieto, one of LVEJO's organizers.

"No transit, no clean air, no Olympics!" shouted members of LVEJO at a press conference outside Mayor Richard M. Daley's office in City Hall. They demanded that the coal plants be replaced with renewable energy job-training centers and alternative energy producers more in line with Chicago's turn toward a green economy.

On November 3, 2007, over 1,400 events were staged across the United States as part of the Step It Up! Campaign initiated by writer Bill McKibben. McKibben had written the first general-audience book on global warming, *The End of Nature*. While teaching at Middlebury College in Vermont, he mentored a group of undergraduates who used a host of social-networking Internet tools to publicize the first Step It Up! day, in April 2007, calling attention to the need for an 80 percent reduction in carbon emissions by 2050. The second Step It Up! day featured the two themes "Green Jobs Now" and "No New Coal." One of the biggest events took place in New Orleans, where presidential candidate John Edwards led hundreds of marchers to the Superdome. There they arranged themselves to form the words "NO NEW COAL," photographed by a helicopter hovering above.

Everywhere I looked around the country, different versions of the same story were playing out. Coal was under attack, and coal projects were being slapped down by regulators and judges. Perhaps even more significant were the large numbers of plants being quietly abandoned by their sponsors. Like an iceberg floating into warming waters, Erik Shuster's list of 151 proposed plants was rapidly shrinking, perhaps faster than anyone had recognized so far.

The project I had casually initiated in the spring of 2007 to track down the status of all 151 plants had turned out to be a bigger undertaking than I had originally imagined. To move it forward, I enlisted the help of several part-time researchers: philosophy graduate student Meilin Chinn, journalist Michelle Chandra, and direct action organizer Adrian Wilson. Each of them became adept at digging through press reports, environmental and financial filings, and activist Web sites, then summarizing the various data into succinct status reports. As our database neared completion, the group continued running across cases of coal plants being canceled, abandoned, or placed on hold. I felt that something significant was taking place. Looking for historical parallels, I read up on the history of the anti-nuclear movement. During the 1970s and 1980s, a combination of grassroots protest and deteriorating economic factors had forced utilities to cancel scores of nuclear plants. I was convinced that the anti-coal movement, though lacking the prominence the "No Nukes!" movement, had established even broader roots across American society and could be on a pace to accomplish more. While "No Nukes!" had been concentrated on the East and West coasts the anti-coal movement was growing in every region of the country.

Others were also tracking the phenomenon. Beginning in the summer of 2007, Matt Leonard at Rainforest Action Network (RAN) had periodically released a summary of plant cancellations and other victories entitled "Moving Closer to a Coal Moratorium." In June the list topped a dozen and continued growing. In November a media report published in the *Denver Post* listed ten cancellations of "clean coal" projects alone. At about the same time, a new Web site appeared with the blunt title Coal Plant Deathwatch Map, showing the location of plant cancellations around the country. A month later, an overseas Web site reported that seventeen U.S. coal plants had been canceled in little more than a year.

The report that seventeen plants had been canceled was astonishing, yet I began to think that the actual number would end up being even higher because our own status list showed that in several instances utilities had canceled plants quietly without notifying the press. In November I met with several organizers from RAN, and the discussion turned toward how the various groups in the far-flung anti-coal movement could most readily share the various informational resources on coal plants that they were developing. One such resource was Carbon Monitoring for Action (CARMA), a worldwide database on carbon dioxide emissions created by the Center for Global Development. Another was the Dirty Kilowatts database maintained by the Environmental Integrity Project, which provided information on coal plant emissions of sulfur dioxide, nitrous oxide, mercury, and other pollutants. A third database, maintained by RAN, contained information on financial institution funding of coal mines and power plants.

Among the various databases, the most widely accessed was maintained by the Sierra Club. It incorporated both RAN's and CARMA's data alongside its own summaries of legal challenges to coal plants. The downside of the Sierra list was that only Sierra's own staff could update or expand the information it contained. At the meeting with RAN's organizers, I mentioned that my friend Earl Killian had suggested that a wiki—an online information database with multiperson-editing capability—would not only allow our own working group to post information more efficiently but would also allow general posting of information by other activists. I liked the wiki approach, since the resulting collaboration would transcend the boundaries of any one group, making it ideally suited to the rapid expansion and increasing diversity of the no-coal movement.

Scott Parkin, one of RAN's organizers, gave me the names of several activist wikis and suggested that I contact them. One was SourceWatch, an information clearinghouse sponsored by the Center for Media and Democracy (CMD) in Madison, Wisconsin. I emailed John Stauber, director of CMD, and he immediately wrote back inviting us to merge our status reports on the 151 proposed coal plants into the 35,000-article wiki database that CMD had already built on topics including the public relations industry, Congress, nuclear power, and Big Tobacco.

Stauber's invitation was appealing for a two reasons. First, by piggybacking onto an existing wiki rather than creating a wiki from scratch, we'd save ourselves time and money. Second, SourceWatch had already accumulated a high degree of "Google juice," that is, the tendency for search engines to give high rankings to content in the SourceWatch wiki. This was due to the large number of Web sites that already linked to SourceWatch articles, as well as the denseness of internal linkages among SourceWatch articles. Both factors are judged by Google's engine to be indicators of a Web page's usefulness to someone seeking information.

Stauber and his collaborator Sheldon Rampton had already thought long and hard about the usefulness of wikis for building activist communities and enhancing collaboration among groups. Through their efforts, SourceWatch had developed ways for each topic focus within the wiki to develop its own unique identity and sense of community. To identify

SourceWatch pages on the topic of coal, we settled on the name CoalSwarm to reflect the anarchic diversity of the no-coal movement and designed a suitable "badge" featuring a cluster of bees.

It took just a few weeks for our small crew to convert our database of coal plants into wiki format. Kaethin Prizer, whose experience included working as a project manager at Yahoo and as a professional book editor, spearheaded the effort. Once we had finished moving the coal plant information into the CoalSwarm wiki, we began creating additional wiki articles on power companies, lobbying groups, citizen groups, and protests, as well as on topics such as "clean coal" and "mountaintop removal."

CoalSwarm quickly turned into a popular site for activists, journalists, students, and others to find information on coal, and over the following months the site attracted hundreds of thousands of visits and grew to over 2,000 pages of information. I was particularly pleased that anyone, anywhere, could post information—some posts came from activists as far away as Australia and Europe. In order to create a page or add information to an existing page, the only prerequisite was to create a log-in name. What kept things honest was that, according the rules of SourceWatch, every morsel of information had to be linked to a published source. Activists, journalists, students, policy analysts, or anyone else using CoalSwarm didn't have to take our word for any piece of information on the site. They could click on the footnote and judge the veracity of the data for themselves.

On occasion, we were asked why we didn't simply post the information we were collecting on Wikipedia, the original wiki and by far the largest. An advantage of SourceWatch over

Wikipedia was the team of professional editors employed by CMD to police the site, especially Tasmania-based Bob Burton, who not only coached new contributors to SourceWatch through the finer points of formatting and referencing their reports, but also contributed major reports of his own on international coal topics.

Without the editorial oversight provided by SourceWatch, Wikipedia proves to be a poor tool for muckraking because information on business misdeeds and controversies is often quietly deleted by image-conscious corporate officials and public relations firms. SourceWatch's editors prevent that from happening, making it a more reliable place to build a clearinghouse on controversial industries such as coal or tobacco.

By the time we had completed our database of proposed coal plants, our list of projects that had been canceled, abandoned, or placed on hold during 2007 had grown to fifty-nine. This struck me as a newsworthy number. I called Matt Leonard at RAN, and we agreed to issue a joint press release. Most press releases are a single page; ours broke that rule, running to seven pages, including four pages of footnotes and one page of links. Knowing that our credibility would be ruined if a single piece of incorrect information slipped into our findings, we checked and double-checked.

Within days of the release, the information was popping up on blogs and in online environmental newsletters. Eventually, it was picked up by the mainstream media as well. The utility and coal industries did not like the publicity about plant cancellations. "This is part of a concerted effort to grossly exaggerate opposition to coal-based electricity generation," said Luke Popovich, a spokesman for the National Mining Association, in an interview with environmental reporter

Steve James. Popovich complained that environmental groups were on a "jihad" and were "exaggerating anecdotal evidence to conclude that coal is on the way out."

But the evidence was not anecdotal. Each cancellation was well documented, and the press release made no attempt to overstate the role of the anti-coal movement in the cancellations.* In fact, we had taken the industry at its word about the reasons plants were being canceled. Among the fifty-nine derailed projects, regulators, courts, or local authorities had rejected only fifteen outright. The sponsors themselves had voluntarily scrapped the majority—forty-four projects. Reasons cited by companies for abandoning plants included rising construction costs, insufficient financing or failure to receive hoped-for government grants, lowered estimates of demand, concerns about future carbon regulations, and competition from renewable power sources, especially wind and solar thermal. Only rarely did utility companies credit opposition by citizen groups as a factor.

While Popovich and other industry spokespeople sought to dispel any sense that King Coal was in trouble, it was hard to dispute the fact that something significant was afoot. Lester Brown, chairman of Earth Policy Institute, wrote:

> What began as a few local ripples of resistance to coal-fired power is quickly evolving into a national tidal wave of grassroots opposition from environmental, health, farm, and community organizations and

---

* Our specific results were challenged in one case: NRG Energy's proposed Huntley Generating Station in Tonawanda, New York. Company officials told reporters that the project had not been placed on hold, as we had reported, but was actually still progressing. A New York assemblyman involved in seeking funding for the plant paraphrased Mark Twain: "Reports of this project's demise are greatly exaggerated." But three months later, NRG admitted that the Huntley Plant was facing a $430 million shortfall, and by midsummer the company had officially announced that it was abandoning the plant. Our report was premature but correct.

a fast-growing number of state governments. The public at large is turning against coal. In a September 2007 national poll by the Opinion Research Corporation about which electricity source people would prefer, only 3 percent chose coal.

Coal's future is also suffering as Wall Street turns its back on the industry. In July 2007 Citigroup downgraded coal company stocks across the board and recommended that its clients switch to other energy stocks. In January 2008 Merrill Lynch also downgraded coal stocks. In early February 2008 investment banks Morgan Stanley, Citi, and J.P. Morgan Chase announced that any future lending for coal-fired power would be contingent on the utilities demonstrating that the plants would be economically viable with the higher costs associated with future federal restrictions on carbon emissions. On February 13, Bank of America announced it would follow suit.

On both sides of the fight over coal, the perception of how financiers viewed new coal plants was becoming a central strategic concern. Rainforest Action Network had long recognized the key role of Wall Street in major energy decisions, and it had made bank policy a direct target of action. So had the Union of Concerned Scientists, which had released a significant body of research pointing out the risks of investing in coal plants. The problems associated with financing coal called to mind the earlier meltdown of nuclear power—not just the physical meltdowns at Three Mile Island and Chernobyl, but also the widespread financial meltdowns among utilities that had embarked on major nuclear plant construction projects in the 1970s and 1980s. Across the country, utilities building nukes had gone bankrupt or defaulted on bonds: Public Service of New Hampshire (bankrupt in the wake of problems at its Seabrook Nuclear Power Plant); Long Island Lighting Company (nearly bankrupt due to problems with its Shoreham plant); Consumers Power in Michigan (nearly bankrupt due to cost overruns at its Midland nuclear plant);

Washington Public Power Supply System (largest municipal bond default in U.S. history—$2.25 billion—due to problems with two nuclear plants). In the case of Washington Public Power Supply System, court battles over who would bear the brunt of the financial collapse of the utility were still being fought a quarter century later. The fiscal difficulties that had befallen utilities building nuclear plants had taught utility bondholders and shareholders to be wary of multibillion-dollar projects surrounded by environmental controversy, prone to major construction cost overruns, or vulnerable to increases in operating costs such as would occur if carbon taxes or cap-and-trade programs were enacted.

Taken collectively, all these concerns made the financial community increasingly wary about large coal plants. Wall Street's jitters explained why plant cancellations were becoming an increasingly fraught subject for King Coal, why opponents of coal sought to publicize such cancellations, and why utilities sought to downplay them. The more coal plants were canceled, the greater the risk in the eyes of bond issuers and other financiers of approving such plants. The greater the risk, the higher the risk premiums that utilities would be forced to pay. Utility executives saw the process as dirty pool, a way for opponents of coal to undermine coal's economic advantage. Activists believed that the nominal cheapness of coal had always been an illusion created by coal's ability to foist its public health and environmental damages onto society, so anything that would raise the cost of doing business was a step in the right direction.

One thing the two sides in the war over coal could agree on was that the wave of cancellations had not ended. Less than a year after Erik Shuster released his list of 151 proposed new coal plants, four out of ten of those projects were no longer moving forward. But in many ways the movement against coal was just getting starting.

■

ELEVEN

Unicorns, Leprechauns, Clean Coal

■

EVERYTHING IS BIGGER in Texas. In March 2008 I went to Houston to attend a Coal Moratorium Now! demonstration and conference along with activists from about twenty states. After two days spent meeting with one another and listening to presentations at a Methodist church in the liberal Montrose neighborhood, we joined other activists from Houston and across Texas in a demonstration outside the massive George R. Brown Convention Center, where the Greater Houston Partnership was holding its second annual America's Energy Futures Forum.

We were on the Avenida de las Americas, and across the street the convention center loomed and stretched out in both directions, making us feel like passengers in a rowboat next to the *Titanic*. Hillary Clinton had come to town, the only candidate to appear before the audience of energy bigwigs. John Edwards had already dropped out of the race. Barack Obama apparently had other fish to fry. We gamely shook our signs and tried to make ourselves heard over the din from the river of cars that separated us from the convention center. A handful of reporters, flipping their steno pads, interviewed protesters. A single TV camera scanned the protest. We'd make it onto the local evening news—maybe.

As I often do at such moments, I felt the futility of it all. I'd seen protests even a thousand times larger disappear through the shrinking magic of the media into a brief story on page 8 of the daily newspaper. Did our comparatively modest turnout amount to more than a blip? How could we be a threat to anyone?

I recalled Gandhi's reassurance: "First they ignore you, then they laugh at you, then they fight you, then you win." That sounded nice, but were we really on the right track? How long would it take?

And yet, there was that number: fifty-nine coal plants canceled, abandoned, or placed on hold during 2007, plus five more in January and February of 2008: a total of sixty-four coal plants canceled in just fourteen months.

Of course, the anti-coal movement could not claim to be the only reason these plants had been stopped. Typically it was a combination: bad economics plus a good shove by activists. But the progress was undeniable. Like an overweight football player huffing and sweating his way through summer training camp, the American economy was trying to shed its fossil fuel addiction and switch to cleaner technologies. Change was happening, even in hydrocarbon-happy Texas. After passing a renewable portfolio standard in 1999, the state's wind power capacity had quadrupled, and Texas now led the nation in wind. It also boasted the largest assembly of wind turbines in the world, the Horse Hollow Wind Energy Center, with 421 massive GE and Siemens turbines spread across forty-seven thousand acres near Abilene. If that could happen here in the citadel of fossil energy, then anything was possible. Change works in mysterious ways. Maybe our prospects weren't so bad. King Coal's planned expansion, despite all the money and political clout that had been poured into moving it forward, was spinning its wheels in the mud of bad economics and mounting opposition. General sentiment appeared to be on our side, as evidenced by a poll released the previous October that showed 75 percent of the public supporting a five-year moratorium on coal plants and increased investment in alternatives like solar, wind, and efficiency measures.

Of course, the political strategists at the National Mining Association and other lobbying groups were neither fools nor quitters. I thought of Bob Henrie, one of the coal industry's senior flacks and political strategists. As the chief of staff for the House Mining and Mines Subcommittee, he'd been at the center of national policy making. He'd also been on the front lines during the worst kind of PR disaster a spokesman ever has to deal with: one in which the company is accused of negligently killing its own employees. In 1984 Henrie had been the flack for the Emery Mining Corporation in Utah following the Wilberg Mine fire, where twenty-seven miners lost their lives in a mine shaft so deep that it took rescuers over a year to dig their way down to the victims' bodies. Through all that time, Henrie had gamely represented Emery, denying charges by workers in the relief-and-rescue effort that at the time of the deadly accident the company had been pushing crews to set a new twenty-four-hour production record for longwall mining. Recently, Henrie had been in the news again, this time after a collapse in the Crandall Canyon mine in Utah buried six miners, and after three rescuers were killed ten days later.

Coal executives obviously trusted Henrie to handle a crisis, and now newspapers reported that the National Mining Association had hired him to develop a new pro-coal advertising and media campaign. Henrie seemed to relish the prospect of helping an unpopular client fight its way out of a corner. He told the *Tribune*, "The advocates of coal haven't had a lot to advocate for. People have a mindset to build a case against coal, rather than for coal. It's our job to keep coal at the table. It's not there now."

The plan that Henrie and the other coal industry strategists developed in early 2008 was a clever one: focus on the presidential primaries. The strategy made sense because it not only gave King Coal the chance to ride the media road show that moved across the country with the candidates—from New Hampshire to Iowa to South Carolina and on—but it also allowed coal supporters to put the candidates on the spot about coal in key states where the sensitive issue of coal worker jobs could make or break an election.

A press release from the industry group American Coalition for Clean Coal Electricity (ACCCE) summed up the strategy: "Presidential Race Runs through the Heart of Coal County, and the Candidates Recognize That Political Reality."

A year after the coal industry strategy began to unfold, memos leaked to the press confirmed numerous details of the plan. But none of it had ever been particularly secret. One key component was the funding of primary debates, especially in states such as Nevada and Florida where numerous coal plant proposals were under consideration. A second piece of the strategy was the "I believe" media campaign that touted "clean coal" through ads on TV, radio, billboards, and the Web. Lavishly produced by R&R Partners, the ad agency behind the

"What Goes On in Vegas Stays in Vegas" campaign, the ads displayed a soft-focus vision of coal as a benign and beneficial mainstay of modern life, conveniently ducking any specifics. Was "clean coal" a current reality, a near-term prospect, or a rosy-hued vision of the future? Somehow the ads implied that the answer could be: "All three."

According to Gristmill blogger David Roberts, the entire "clean coal" notion rested on a deliberate use of ambiguity:

> The "clean coal" PR people are running a scam. Thing is, it's an obvious scam—easily exposed, easily debunked. Just because it's obvious, though, doesn't mean the media won't fall for it. Indeed, the entire "clean coal" propaganda push is premised on the media's gullibility.
>
> Here's the scam: They leave the definition of "clean coal" deliberately ambiguous. As ACCCE spokesman Joe Lucas said on NPR the other day, "clean coal is an evolutionary term." By "evolutionary," of course, he means, "whatever we need it to mean at the moment." If one meaning is attacked, they subtly shift to another meaning.

Certainly the vision of coal plants that could economically bury all their pollutants safely and permanently far beneath the ground was an attractive idea, but according to a detailed report released by Greenpeace, "the earliest possible deployment of carbon capture and storage at utility scale is not expected before 2030." In addition to the high projected costs of the process, numerous technical, legal, and institutional problems remained unsolved. One nagging issue had to do with enforcement. Given that running carbon capture equipment would require at least a quarter of a plant's output, what was to prevent plant operators—especially in countries with poor regulatory standards—from cheating in order to maximize power output? Even if such issues could be dealt with, costs for producing electricity from "clean coal" were projected to be significantly higher than costs of cleaner alternatives such as efficiency measures, solar thermal power, and wind power. But if the cleaner alternatives were also cheaper, why bother with coal at all?

As for the notion that clean coal—or even relatively clean coal—was already becoming a reality for new coal plants, here's a tally of what one proposed coal plant, the 250-megawatt Highwood Power Project in Montana, characterized by its sponsor as the "cleanest in the country," would release each year, according to its draft air quality permit:

- Three million tons of carbon dioxide, the most important greenhouse gas, an amount equivalent to chopping down 130 million trees.

- 443 tons of sulfur dioxide, which causes acid rain and forms small airborne particles that produce lung damage, heart disease, and other illnesses. Fine particulates from power plants (both emitted directly and formed from sulfur dioxide) are responsible for 550,000 asthma attacks, 38,000 nonfatal heart attacks, and other cardiopulmonary disorders. They also cause 24,000 premature deaths each year in the United States, the average mortality resulting in fourteen years of lost life.

- 944 tons of nitrogen oxide (NOx), equivalent to 50,000 late-model cars. NOx leads to formation of smog, which inflames lung tissue and increases susceptibility to respiratory illness.

- 44 tons of hydrocarbons, which contribute to smog formation.

- 1,177 tons of carbon monoxide, which causes headaches and places additional stress on people with heart disease.

- 40 pounds of mercury. One-seventieth of a teaspoon of mercury deposited in a twenty-five-acre lake can make the fish unsafe to eat. Over 600,000 babies are born annually to women with unsafe levels of mercury in their bodies, leading to learning disabilities, brain damage, neurological disorders, and other health effects.

■ 366 tons of particulate matter, a catch-all category that includes metals such as arsenic, beryllium, cadmium, manganese, and 560 pounds of lead. These toxic metals can accumulate in human and animal tissue and cause serious health problems, including mental retardation, developmental disorders, and damage to the nervous system. Arsenic leads to cancer in 1 out of 100 people who drink water containing a mere 50 parts per billion.

It's one thing to read such a list. It's another to experience the actual pollution on the ground, and the No New Coal Plants listserve provided occasional reminders of that reality, such as the following report by Elisa Young, who lived near several coal plants in Meigs County, Ohio: "Three out of three guests staying at my farm this week suffered from breathing problems—all three wheezing, one who had no history of asthma, and I found it very hard to breathe, feeling lethargic when it was nice outside and should have been a good day to get work done."

For those wrestling with the possibility that a large power plant may be sited in their community, one challenge is to get others in the community merely to begin imagining how much such a facility will change the fabric of daily life. On a trip to meet with activists in upstate New York, I visited the bucolic town of Jamesville, where City Councilwoman Vicki Baker showed me the location that had been selected by a New York City–based entrepreneur, Adam Victor, to build one of the world's largest coal gasification plants. We walked along a little-used railroad spur line, strewn with lumps of bituminous coal. Next to the tracks was a fence partially overgrown with brambles, and close at hand were a number of houses with well-tended yards. It was hard to believe that this small town had been considered an appropriate site for a facility the size of an oil refinery.

Vicki recalled the intense local organizing that had followed the announcement, led by her group Jamesville Positive Action Committee (JAM-PAC). At public meetings, residents questioned how such a megafacility could operate without endangering an elementary school located a stone's throw away. The sheer size of the plant, designed to turn one hundred train cars of Pennsylvania or West Virginia coal into methane gas every day, was hard for people to grasp.

Liz Curly, parent of a seven-year-old boy, told one meeting, "My concern is the fact that refineries have accidents all the time. We're dealing with methane gas, which is explosive. Evacuation would be troublesome. Where my son plays and learns should be the safest place."

Although the developers insisted that the project would be safe, residents already had experience with a coal ash storage facility on the same site, and they had experienced frequent releases of ash despite company assurances to the contrary. They also lived close to a hazardous waste incinerator and had become disillusioned with that project's insistence that it, too, was safely operated.

As we left lunch at a diner in Jamesville, Vicki suddenly pointed at a sudden puff of what looked like smoke rising from the coal ash storage facility.

"Is that normal?" I asked.

"It's not allowed under the permit," she said. "But it happens."

Vicki pulled out her cell phone, called an enforcement officer with the New York environmental quality department, and reported the release. From the familiar tone of the conversation, it sounded like the two had spoken many times. Since she was a member of the local city government, her call could not easily be ignored.

It's likely that when Adam Victor drew up his plans to site a giant gasification plant in Jamesville, he failed to foresee the sort of organized opposition that Vicki Baker and others in the town would put together. Developers typically benefit from the inherent boosterism of small towns. Local politicians tend to seize on promises of jobs—any jobs—to the exclusion of all other concerns. By the time opposition to a large project such as a coal plant begins to find its feet, city officials have already formed relationships with company officials, and the wheels of various permitting processes are turning. Local activists then face the twin challenges of trying to gain access to information while at the same time slowing a train that has already begun to pull out of the station.

Here in Jamesville, Vicki Baker and other opponents of the gasification project had managed to scramble fast enough to get traction before it was too late to make a difference. The group was politically experienced, and within short order a slate of anti-project candidates had ousted pro-project members of the town council. "Stop the Coal Plant" lawn signs sprouted throughout the town, especially after it was revealed that the project would include train cars containing sulfuric acid and mercury, that the plant would include a 110-foot flare tower, and that noise from the plant would be significant. As the bare facts of the project emerged, local boosters ran for cover and support for the project evaporated.

Vicki told me that we didn't have time to wait for the state enforcement officer to arrive, because she had scheduled a meeting with residents in the town of Scriba, near the Lake Ontario resort city of Oswego. That was the fallback location selected by Adam Victor for the coal gasification plant after noting the level of community opposition in Jamesville. Now the wheels of local organizing were beginning to turn at the new site, this time under the leadership of engineering professor Dr. Kestas Bendinskas.

The dismissive term for the sort of meetings-in-the-living-room activism that generally confronts developers is NIMBY: Not in My Back Yard. The process often plays out like a game of whack-a-mole. When citizens in one community turn out to be excessively feisty, developers pick up stakes and find a more amenable location. If Kestas and others were sufficiently tough, resourceful, and organized, perhaps they could send Adam Victor down the road to yet a third town, one where people were less empowered. In the end the project might land in a poorer, less cohesive community.

Whatever the realities of coal at the local community level, the coal industry was looking at buying its way to acceptance on a vastly larger scale. According to the *Washington Post*, ACCCE made a $35 million commitment to the "clean coal" advertising campaign aimed at key primary and caucus states in the 2008 presidential campaign. The same newspaper reported that the National Mining Association had increased its 2008 lobbying budget by 20 percent from the previous year.

On top of its advertising artillery, the coal industry deployed paid outreach workers to attend rallies and debates throughout the primary states. With their "clean coal" hats, shirts, and signs, the outreach workers were never far from view as the candidates made their stump speeches. A goal of the campaign was to get the candidates on record in support of governmental investments in "clean coal" technology, and soon both the Obama and the McCain campaigns were swearing fealty to the clean coal message. In one widely quoted remark, Barack Obama told a crowd in West Virginia: "This is America. We figured out how to put a man on the moon in ten years. You can't tell me we can't figure out how to burn coal that we mine right here in the United States of America and make it work."

Whether the coal industry expenditure would ultimately pay off was more debatable. Early in the primary season, Obama met with editors at the *San Francisco Chronicle*, and coal and climate change was a major point of discussion. Obama's remarks were blunt, revealing a more complex viewpoint than the one he had expressed in West Virginia. Obama said:

Let me sort of describe my overall policy.

What I've said is that we would put a cap-and-trade system in place that is as aggressive, if not more aggressive, than anybody else's out there.

I was the first to call for a 100 percent auction on the cap-and-trade system, which means that every unit of carbon or greenhouse gases emitted would be charged to the polluter. That will create a market in which whatever technologies are out there that are being presented, whatever power plants that are being built, that they would have to meet the rigors of that market and the ratcheted down caps that are being placed, imposed every year.

So if somebody wants to build a coal-powered plant, they can; it's just that it will bankrupt them because they're going to be charged a huge sum for all that greenhouse gas that's being emitted.

That will also generate billions of dollars that we can invest in solar, wind, biodiesel and other alternative energy approaches.

The only thing I've said with respect to coal, I haven't been some coal booster. What I have said is that [it's wrong] for us to take coal off the table as an ideological matter as opposed to saying if technology allows us to use coal in a clean way, we should pursue it.

The interview was not released until several days before the election. The McCain campaign immediately began broadcasting it in coal states such as West Virginia, Illinois, and Ohio, but the last-minute push to paint Obama as an enemy of coal failed to change the ultimate outcome in any of the states.

Overall, the coal industry's "clean coal" campaign and its focus on the presidential campaign revealed the industry's strengths and weaknesses. In key mining states, such as West Virginia, Wyoming, and North Dakota, the industry had always enjoyed tremendous clout. Senators such as Jay Rockefeller of West Virginia had long played the role of pitchmen for coal. Through his chairmanship of the Commerce, Science, and Transportation Committee, Rockefeller in particular was able to bring home large subsidies for coal projects in West Virginia.

Outside the frontline coal states, the industry lacked a strong base, and in numerous other states—Florida, California, Maine, Washington, Montana, Kansas, Colorado, Texas, and Minnesota—coal's opponents were winning an increasing number of skirmishes. Moreover, the movement was attracting increasing support as nationally based efforts like Al Gore's We Campaign, Working Assets and CREDO Mobile, Coop America, Citizens Lead for Energy Action Now (CLEAN), the National Parks Conservation Association, and scores of others recruited thousands of people to lobby against coal plants and mountaintop removal mining.

Against its growing array of foes, King Coal continued spending heavily to promote its "clean coal" message. But that strategy was not without its risks. Having taken such strenuous measures to brand itself in the public mind as "clean," what would happen if uncomfortable realities intruded on that spanking-clean image, such as massive coal waste spills or other environmental mishaps? It was a question being answered on a daily basis in the mountains and valleys of Appalachia, if only the rest of the country could be persuaded to take a look.

■

TWELVE

War Against the Mountains

■

NOWHERE IN AMERICA was the absurdity of the coal industry's "clean coal" PR campaign more blatantly obvious than in Appalachia, where mountaintop removal mining had turned large parts of the most beautiful forested areas in the country into a wasteland. Word of the destructive practices was getting out. Not that many years earlier, when anti-coal groups in different parts of the country had worked more or less in isolation, citizens in Appalachia who attempted to oppose mining companies did so with little outside support. But increasingly the various strands of the movement were discovering one another, and mountaintop removal had emerged as a cause célèbre.

In 2003 tenth-generation Appalachian and self-proclaimed "endangered hillbilly" Judy Bonds won the Goldman Award, the environmental movement's equivalent of the Nobel Prize, for her work with Coal River Mountain Watch. Shows like *Frontline* were beginning to air documentaries about mountaintop removal, and people across the United States were finally seeing firsthand the dirty war being waged against the mountains and the people of Appalachia.

Although coal is mined and consumed in many parts of the United States, Appalachia's long tradition of coal mining has deeply formed the character of the region, and the struggles surrounding coal have always burned with particular intensity and even violence. I knew this from my own family history, since my mother was born in Dundon, a company-owned town in Clay County, West Virginia. In those days Dundon was not accessible by road; the only way in or out was to ride the train or walk along the tracks. According to family lore, my grandfather, a Presbyterian preacher, had been run out of town by the coal company for preaching pro-union sermons. He may have gotten off relatively easy. According to historical accounts of the period, union organizers in many parts of West Virginia were not only prohibited from holding meetings or entering miners' homes, many were also arrested, beaten, and even killed.

On a Delta Airlines flight into the TriCities Airport near Blountville, Tennessee, I looked out the window into the geographical nexus where five states—Virginia, Kentucky, Tennessee, North Carolina, and West Virginia—come together. Extending along a southwest/northeast axis, these same mountains stretched like rumpled corduroy from Alabama to New York. Even from twenty thousand feet, you could see how the landscape had historically fragmented the region, and why solidarity has meant so much to generations of union and environmental organizers. On top of the physical topography, a spiderweb of towns, roads, and rails traced the strands of civilization.

I was on my way to a strategy session in Abingdon, Virginia, hosted by the Alliance for Appalachia, that aimed to build tighter working relationships between Appalachian groups and activists from outside the region. Driving along Interstate 81 out of Blountville, I marveled at the contrast between the peaceful landscape, its colors still muted by winter, and the brutal history of these mountain states, where at times labor conflicts had blossomed into full-scale rebellion.

In 1921, four years before my mother was born, approximately thirteen thousand West Virginia coal miners participated in a series of gun battles against a private force of two thousand hired guns underwritten by the Logan County Coal Operators Association. The trigger for the uprising was the murder of Sid Hatfield, the union-friendly police chief in the hamlet of Matewan, by agents of the Baldwin-Felts detective agency. Some miners commandeered a Chesapeake & Ohio freight train. After initial skirmishes, President Warren Harding ordered federal troops into the conflict, and Army Martin MB-1 airplanes dropped bombs on the miners. By the time the federal troops arrived, as many as thirty detectives and one hundred miners had lost their lives. Following the battle, 985 miners were indicted for "murder, conspiracy to commit murder, accessory to murder, and treason against the State of West Virginia." Short term, the outcome seemed to be an overwhelming victory for management. Federal judges backed up the owners with blanket injunctions barring union organizing throughout several counties. United Mine Workers (UMW) membership in the state plummeted from more than fifty thousand miners to approximately ten thousand. Not until 1935, after the election of Franklin Delano Roosevelt, did the UMW regroup and fully organize in southern West Virginia.

As William Faulkner wrote in reference to his native Mississippi, "The past is not dead. In fact, it's not even past." Today, company-sponsored violence continues in Appalachia, though the target has shifted. There's harassment and targeting of anti-mining activists. There's also the violence of mining itself, carried out with massive machinery and three million pounds of explosives each day.

Strip mining in Appalachia dates to the 1950s, and some of today's grassroots groups, such as Save Our Cumberland Mountains, date to the early 1970s. Early efforts focused on enacting legislation that would abolish strip mining altogether, but the comprehensive federal law that was finally enacted in 1977, the Surface Mining Control and Reclamation Act, lacked teeth, and the teeth that it did have were repeatedly blunted by weak enforcement. Passage of the act dissipated the energy of the movement while not solving the problem. In the wake of that legislative fiasco, much of the movement collapsed, only to begin reviving again in response to an even more devastating form of mining known as mountaintop removal.

Measured against the long history of mining in the Appalachia, this new way of getting at coal is a recent innovation. The first such operation began in 1970 when Cannelton Industries began blasting the top off Bullpush Mountain in Fayette County, West Virginia, pushing the rubble into adjacent valleys to expose the underlying coal. Cannelton said it wanted to build a new town on the site, including churches, schools, stores, and a hospital. No town was ever actually built, and the site remains desolate.

Once the topography itself is destroyed, all other aspects of a natural area follow into oblivion, from streams and underground aquifers to biotic communities. The effects are especially profound considering that the Appalachian region hosts one of the most diverse temperate forests in the world. Having escaped the six-mile-thick glaciers that once covered most of the Northeast, the trees of Appalachia provided the initial seeds for the plant species that recolonized the hundreds of thousands of square miles of terrain to the north. The irony is that mountaintop removal mining is a process as destructive as glaciation, yet the seeds used to replant the spoil piles are *Lespedeza cuneata*, an invasive species of legume introduced from Asia. The result, like a shadow of stubble appearing on the face of a recently deceased body, is a parody of an ecosystem.

During the 1980s the extent of mountaintop removal remained relatively modest, affecting fifteen square miles during the course of the entire decade. The pace increased during the 1990s and then accelerated after a fateful meeting in 1999. That year, an 88-year-old West Virginia mine operator named James "Buck" Harless flew to Austin, Texas, for lunch with then-governor George W. Bush. The goal of the luncheon was to sign Harless up as a Bush Ranger, one of hundreds of wealthy backers who each committed to raising $100,000 from smaller donors. But Harless had an agenda of his own. Although West Virginia had a long history as a Democratic stronghold and had backed Clinton in the previous presidential election, he believed he could deliver the state to Bush, thereby winning his industry a front-row seat in the coming administration. Toward that end, he poured himself into fundraising and campaigning, reaching out to mine workers with the message that the election of Gore would mean disaster for West Virginia coal. In the end those efforts proved successful, and West Virginia moved into the Republican column. At the close of the 2000 campaign, all eyes were on the disputed results in Florida, which the Supreme Court finally resolved in favor of Bush. But Florida would have been irrelevant if West Virginia's five electoral votes had stayed Democratic.

After the election, William Raney, the head of the West Virginia Coal Association, told the association's members that they could now expect a "payback" from the new administration. Within months, that reward had arrived in the form of a small but fateful change in the definition of mountaintop removal debris, secured by Deputy Interior Secretary Stephen Griles. Instead of calling the debris "waste," which would be prohibited from entering fresh waterways under the Clean Water Act, Griles directed that mine debris be regarded as "fill," an acceptable category. The result was a green light to mountaintop removal, which immediately accelerated. In 2002 alone, permits were issued covering twenty square miles.

The explosion of mountaintop removal mining under the Bush administration did not go unchallenged. Throughout Appalachia, citizen groups fought back with lobbying efforts, grassroots organizing and education, media outreach, litigation, regulatory input, marches, and direct action protests. They also sought to expand their connections beyond the region through projects like Dave Cooper's Mountaintop Removal Road Show and events like the one I was attending in Abingdon. Here, in a large lecture hall at the Southwest Virginia Higher Education Center, activists from half a dozen Appalachian states were gathering with California-based organizers from Rainforest Action Network (RAN) to discuss the adoption of a "market campaign" strategy to stop mountaintop removal.

RAN had previously used market campaign tactics to protect old-growth forests, winning concessions on paper and lumber purchasing policies from Burger King, McDonald's, Mitsubishi, Home Depot, and other corporations. As RAN's Jennifer Krill explained to the attendees at the Abingdon meeting, the leverage developed in each such campaign grew directly out of the value that companies cultivate through their brands and corporate image building. By drawing public attention to destructive practices like old-growth logging or mountaintop removal mining, market campaigns threaten to undermine the value of a company's image. Krill reported on the "Carbon Principles" that three leading Wall Street banks—Citibank, JPMorgan Chase, and Morgan Stanley—had established the previous month. In a vaguely worded statement, the banks had promised to "pursue cost-effective energy efficiency, renewable energy and other low carbon alternatives to conventional generation." Anti-coal groups had seen the statement as inadequate, but it showed they had won the attention of the banks and provided a platform for exerting further pressure.

A key step toward building national awareness of mountain-top removal was the creation of Mountain Justice Summer, an effort modeled after the Freedom Summer campaign to register Black voters in the Deep South in 1964 and the Redwood Summer campaign to block old-growth logging in California in 1990. In the first Mountain Justice Summer, in 2005, Judy Bonds and other members of Coal River Mountain Watch invited about fifty young activists to West Virginia to reinforce protests against a coal silo and a 2.8-billion-gallon coal-slurry reservoir operated in Sundial, West Virginia, by a subsidiary of Massey Energy. The coal silo, which emitted dangerous coal dust on a regular basis, was situated adjacent to the Marsh Fork Elementary School, and the coal-slurry reservoir was less than a quarter mile uphill from the school. The failure in 1972 of a similar impoundment, the Buffalo Creek Dam near Charleston, had resulted in the death of 120 people when a river of sludge crashed through several mountain hamlets. Residents of Sundial feared that a repeat of the Buffalo Creek disaster could destroy the school and the community.

Throughout the summer of 2005, the Coal River Mountain

Watch and Mountain Justice Summer activists staged a series of rallies, marches, sit-ins, and other protests at Sundial, at Massey's headquarters in Virginia, and at the capitol building and the governor's office in Charleston. The following summer, the protests resumed.

Coal River Valley resident Ed Wiley, whose granddaughter Kayla Taylor attended Marsh Fork, knew the 385-foot-tall impoundment dam better than most, having been part of the construction crew. Wiley happened to have a knack for publicity. Around the community, he passed the hat for the Pennies of Promise campaign, an effort to raise money to move the school to a safer location. By offering the collected pennies as a symbolic down payment, Wiley hoped to shame officials into allocating the necessary funds. Just to ensure that the gesture was noticed, he decided to personally deliver the pennies to Senator Robert Byrd's office in Washington, D.C.—on foot. On August 2, 2006, Wiley began his forty-day walk, trekking along the side of the highway and gradually attracting a widening circle of attention from the press.

Across the region, other Appalachian leaders were also mastering the art of media, and a key part of that mastery had to do with weaving protest together with celebration of mountain heritage. To filmmakers like Michael O'Connell, the underdogs-versus-overlords storyline was compelling. In O'Connell's film *Mountaintop Removal*, a tired but determined Ed Wiley makes his way inexorably toward the U.S. Capitol, carrying a heavy flagpole on his shoulder cushioned by a folded towel. It is an iconic image of grassroots pride and resistance.

In a similar vein, groups like Kentuckians for the Commonwealth (KFTC) made the slogan "I Love Mountains" synonymous with opposition to King Coal. At KFTC's annual Valentine's

Day rally on the capitol steps in Frankfurt, mountain music by Clack Mountain String Band, Public Outcry, and Randy Wilson kept the crowd in high spirits as speakers prepped the attendees to lobby state legislators.

Not surprisingly, given the ubiquity of religion in Appalachian culture, activists were working at every level from local church discussion groups to interdenominational organizations formed to educate fellow Christians about mountaintop removal. At the Abingdon conference, activist Maria Gunnoe handed me a DVD entitled *Mountain Mourning*, which juxtaposed images of beautiful mountain scenery and biblical verses pertaining to the sanctity of nature with horrific photos of mining destruction. The DVD had been created by the group Christians for the Mountains, organized in Charleston, West Virginia, in ~~2005~~. 2007

Of all the creative outreach that Appalachian activists were developing, one of the most effective was the volunteer pilot association Southwings, which took journalists, politicians, celebrities, and benefactors on tours over mountaintop removal sites. The flights and the films posted online by pilots provided a far more revealing view of the devastation than could readily be seen from roads and other locations accessible to the public. Again and again, the flights had succeeded in transforming lukewarm opponents—and at times even some mining supporters—into advocates for banning the practice.

At the Abingdon conference a plan circulated among activists both in the general session and in breakout brainstorms. Since the necessity to provide jobs in an economically troubled area was consistently used as the rationale for continued exploitation of coal, why not turn the issue to advantage by showing how an alternative use of the same land could provide a greater amount of more durable employment? By the end of 2008, a *in his 50's !*

small group spearheaded by Coal River Mountain Watch and two young activists, Rory McIlmoil and Lorelei Scarbro, had developed a comprehensive proposal to build a wind farm rather than a mine on Coal River Mountain. The gist of the proposal rested on the fact that high-elevation areas experience the strongest winds, a resource that would be destroyed if mining flattened the mountain. A study by the environmental consulting firm Downstream Strategies fleshed out the details of the wind alternative. While coal mining would provide roughly a hundred jobs versus fifty jobs for a wind farm, the coal jobs would disappear after fourteen years, whereas the wind jobs would continue indefinitely.

Lenny Kohm, campaign director of Appalachian Voices, was quarterbacking yet another initiative, the Clean Water Protection Act. The proposed act, consisting of a single succinct paragraph, would amend the Federal Water Pollution Control Act to clarify that toxic rubble created by mountaintop removal mining cannot be defined as "fill material" and dumped into the headwater streams of Appalachia. The strategy was to build a list of congressional cosponsors. In 2002 thirty-six congresspeople signed on as cosponsors; in 2003 the number climbed to sixty-four. By 2008 there were 152 cosponsors. Each year, Mountaintop Removal Week provided a focal point for grassroots activists to converge on D.C. and promote the legislation. In 2007 over one hundred citizen lobbyists arrived from nineteen states. In addition to Appalachian Voices, the groups pushing the Clean Water Protection Act included the Appalachian Citizens Law Center, Appalshop, Coal River Mountain Watch, Heartwood, Kentuckians for the Commonwealth, Mountain Association for Community Economic Development, Ohio Valley Environmental Coalition, Save Our Cumberland Mountains, Sierra Club Environmental

Justice Program, Southern Appalachian Mountain Stewards, Southwings, and West Virginia Highlands Conservancy.

Meanwhile, awareness of mountaintop removal mining was growing in urban areas. An early tool in developing this awareness was the "Are You Connected?" campaign, a Web-based tool that allowed residents of cities far removed from Appalachia to find out whether their local power company was using coal mined by mountaintop removal simply by entering their zip code into a computer. Ingeniously designed, the campaign provided activists with "widgets" that could be embedded into blog pages. In July 2008 two Manhattanites organized the first annual New York Loves Mountains festival, including presentations by activists and legislators, music by New York and Appalachian bands at the Jalopy Theater, and a new play written by Sarah Moon. The event led to the formation of an ongoing group of New Yorkers linked up with Appalachian activists.

Further evidence that mountaintop removal had become a top priority not just for regional activists but for the nationwide environmental movement came in April 2009, when Maria Gunnoe became the second West Virginia anti-coal activist to win the Goldman Award.

Throughout the spring of 2009, hopes repeatedly rose and fell that the Obama administration would take strong action to outlaw mountaintop removal once and for all. In a public letter to Obama, Bo Webb of Coal River Mountain Watch expressed the desperation of the movement. Webb wrote:

> As I write this letter, I brace myself for another round of nerve-wracking explosives being detonated above my home in the mountains of West Virginia. My family and I, like many American citizens in Appalachia, are living in a state of terror. Like sitting ducks waiting to be buried in an avalanche of mountain waste or crushed by a falling boulder, we are trapped in a war zone within our own country.

In 1968, I served my country in Vietnam, as part of the 1st Battalion 12th Marines, 3rd Marine Division. As you know, Appalachians have never failed to serve our country; our mountain riflemen stood with George Washington at the surrender of the British in Yorktown. West Virginia provided more per capita soldiers for the Union during the Civil War than any other state; we have given our blood for every war since.

We have also given our blood for the burden of coal in these mountains. My uncle died in the underground mines at the age of 17; another uncle was paralyzed from an accident. My Dad worked in an underground mine. Many in my family have suffered from black lung disease.

These mountains are our home. My family roots are deep in these mountains. We homesteaded this area in the 1820s. This is where I was born. This is where I will die.

Mr. President, when I heard you talk during your campaign stops it made me feel like there was hope for Peachtree and the Coal River Valley of West Virginia. Hope for me and my family.

I beg you to relight our flame of hope and honor, and immediately stop the coal companies from blasting so near our homes and endangering our lives. As you have said, we must find another way than blowing off the tops of our mountains. We must end mountaintop removal.

Despite the hopes that Obama would take action on mountaintop removal, those versed in the history of the region, remained cautious, recalling how earlier dreams of ending destructive mining had been thwarted. But time was running short. By some estimates, only ten to twenty years of economically minable coal remained in Appalachia. Eventually, mountaintop removal mining would surely end, but when it did, what would be left?

■

The Grandmother Rebellion

■

AT A MEETING IN Washington, D.C., among grassroots anti-coal groups, longtime Appalachian activist Larry Gibson turned to two Navajo women who had come to the gathering as representatives of the Black Mesa Water Coalition.

"What happened to you was the blueprint," said Gibson. "Now it's metastasized all over the country."

Gibson's family has lived on Kayford Mountain in West Virginia since the late 1700s, and more than three hundred of his relatives are buried in the family cemetery that now sits isolated above a devastated landscape. Since 1986 he has watched the destruction of Kayford Mountain while enduring relentless personal harassment. His dogs have been shot; there are bullet holes in the siding of his cabin. But Gibson was right. The assault of the coal industry on Navajo and Hopi country was part of a far longer story of conquest and exploitation whose roots traced deep into the genocidal policies of white expansionism.

The U.S. Army established Fort Defiance and Fort Wingate on Navajo land in 1851, and in 1864 thousands of Navajo people were marched over three hundred miles to southeastern New Mexico. More than two hundred people died during the Long Walk. Four years later, most people returned from the relocation camps, but the experience left indelible scars. In 1882 an executive order by President Chester Arthur created a new Indian reservation consisting of a near-perfect square of land that enclosed one of the richest energy deposits in the world, the Black Mesa coalfield. Federal surveyors had recently assessed the coal deposit, and President Arthur's motive in establishing the reservation was to prevent nearby Mormon settlers from laying claim to the land and its rich deposits under the Desert Lands Act of 1877.

Both Navajo and Hopi people lived within the boundaries of the Black Mesa reservation. In the 1960s the Hopi tribal council approved leasing Black Mesa coal to Peabody Coal Company, as did the Navajo tribal council. But subsequently it came to light that attorney John Boyden, who had handpicked the Hopi council and represented the tribe for thirty years, had secretly worked for Peabody Coal during the time he was officially representing the tribe.

Strip mining began at the Navajo Mine in 1963, and by the mid-1970s the Four Corners region had developed into one of the largest coal complexes in the United States, powering Las Vegas, Los Angeles, Phoenix, and other areas on the southwestern power grid. The second mine on Black Mesa was the Kayenta Mine, supplying the Navajo Generating Station. Royalties and taxes from the mines provided approximately 80 percent of the Hopi general operating budget and 60 percent of the Navajo general fund budget. The power plants were massive in scale. At 2,410 megawatts, the Navajo Generating Station was the fourth largest power plant in the United States. The Four Corners plant was nearly as large. Built in the early 1960s, its plume was seen from space by the Apollo astronauts. Two other plants used Navajo and Hopi coal: the 1,800-megawatt San Juan Plant and the 1,640-megawatt Mohave Generating Station in Nevada, which burned coal shipped 273 miles by slurry pipeline from the Black Mesa Mine.

While the mines and plants generated employment, a common complaint on the reservation was that Navajos and Hopis were filling few of the higher-paying jobs. On Black Mesa, 80 percent of Navajo people still lack running water, and 50 percent of people on the Navajo and Hopi reservations lack electricity, a huge irony given the massive transmission lines overhead. In a 2004 *Los Angeles Times* article, Black Mesa resident Nicole Horseherder said, "Somewhere far away from us, people have no understanding that their demand for cheap electricity, air conditioning and lights 24 hours a day has contributed to the imbalance of this very delicate place."

Reservation physician Marcus Higi testified that he had never seen worse asthma than the cases he found on the Navajo reservation. During four years on the reservation, he had to fly five children to hospitals in order to save their lives. Research reported by the U.S. Geological Survey showed that people living in the Shiprock area, where thermal inversions trapped emissions from two nearby coal plants, were more than five times as likely to experience respiratory complaints as residents of nearby communities. In an area where air quality had once been pristine, the power plants had created smoglike conditions worse than those in congested urban areas.

Erich Fowler, a resident of Kline, Colorado, about thirty miles from the Four Corners plant, testified at EPA hearings that a yellow haze "as bright as daffodils" blocked his view of

Farmington and that at times "the sky begins to look like it's filled with scrambled eggs." The American Lung Association estimated that sixteen thousand people in the region (15 percent of the population) suffer from lung disease probably caused by plant emissions. Each year, the San Juan generating station emits approximately 100 million pounds of sulfur dioxide, 100 million pounds of nitrogen oxides, 6 million pounds of soot, and at least 1,000 pounds of mercury. The Four Corners plant emits 157 million pounds of sulfur dioxide, 122 million pounds of nitrogen oxides, 8 million pounds of soot, and 2,000 pounds of mercury. Even the Grand Canyon was affected: photographs showed its depths obscured by yellowish brown smog.

In addition to air problems, those living in the vicinity of strip mines, mainly farmers and sheep ranchers, suffered from water toxicity or loss of water supplies. Furthering the pressure on water supplies was the annual removal of over a billion gallons of water from the Navajo Aquifer to feed the coal slurry between the Black Mesa Coal Mine and the Mohave Station. Runoff from coal mines supporting the Four Corners and San Juan plants contaminated aquifers with sulfates, leading to the death of livestock. Another hazard to water supplies was 150 million tons of coal combustion waste (containing cadmium, selenium, arsenic, and lead) that had been dumped in the Navajo and San Juan mines.

Alongside the environmental impacts came severe sociological upheaval. In 1974 attorney John Boyden and his coal industry allies pushed legislation through Congress that directed the relocation of fourteen thousand Navajo families. Additional legislation in 1996 required the remaining families to move. It was the largest forced relocation in the United States since the internment of Japanese Americans during World War II. Thayer

Scudder, professor of anthropology at the California Institute of Technology, protested the action to the United Nations, writing, "I believe that the forced relocation of Navajo and Hopi people that followed from the passage in 1974 of Public Law 93-531 is a major violation of these people's human rights. Indeed this forced relocation of over 12,000 Native Americans is one of the worst cases of involuntary community resettlement that I have studied throughout the world over the past 40 years."

Federally appointed Relocation Commissioner Roger Lewis resigned in protest. Lewis said, "I feel that in relocating these elderly people, we are as bad as the Nazis that ran the concentration camps in World War II."

Now, another power plant was being slated for Navajo/ Hopi lands, a 1,500-megawatt facility known as Desert Rock. The project was sponsored by Sithe Global Power, a "merchant power" company that planned to sell the power from the project to utilities in the Southwest. It was backed by the private equity firm Blackstone. Enticed by the promise of a $50 million annual payout to the Navajo Nation, the Tribal Council voted 66–7 in favor of inviting Sithe to build the plant, but the plan quickly ran into strong grassroots opposition.

When I met Dáilan Long, one of the organizers with Diné CARE (Citizens Against Ruining Our Environment), I was struck by the quiet confidence and persuasiveness of someone still in his early twenties. Raised on the Navajo reservation and educated at Dartmouth, Long had returned to help organize against the Desert Rock power project. He was quick to say, however, that his role was that of a supporting player, noting that the leadership of groups like Diné CARE rested in the hands of the elders.

"In our culture," said Long, "you do what the grandmothers tell you."

In the Diné language spoken by the Navajo, or Diné, people, the word *doodá* means simply "no." It's also the first word in the name of another Navajo/Hopi activist group: Doodá Desert Rock. If the group succeeds, the proposed Desert Rock Coal Plant will not be built and the Navajo Nation will not receive an annual payment of $50 million from the plant's sponsors. Yet despite the loss of the promised payment that would result from killing the project, Doodá Desert Rock and other groups opposing the plant appear to enjoy widespread support.

Among dozens of comments about the Desert Rock plant collected by Ecos Consulting, the following were typical:

A rancher/farmer: "I lost five of my female cows and each of them was with an unborn calf during the winter from drinking contaminated water in the mining area. The energy corporation creates hopes and dreams they do not keep. We don't need the power plant and we don't need the coal mine to survive; our people survived for many centuries without any power plants and coal mines so why should we need it now?"

A weaver/rancher: "Two power plants and one more on the way are too many power plants, and I opposed all of them. We are already badly polluted by all kinds of toxics and who is cleaning it up? Nobody. We are sick and most of the people around Four Corners power plant and surrounding areas have numerous health problems. We can even smell the smoke from the smoke stacks in certain temperatures or the way the wind blows."

A rancher: "I oppose another coal-fired power plant. We have already experienced the bad side of Four Corners power plant. We have been there. They lied to the people and all the promises were never fulfilled. Why should we go for any other power plant with the same empty promises?"

A nineteen-year-old student: "Being a Christian, we must LOVE our people and protect what God has provided for us to live with. We should not reject or in any way misuse what God gave us. We need to protect our cultural sites, traditional burial grounds, our holy offering sites, and historical sites by not contaminating the air, water, and land.

Desert Rock power plant will pollute our land and put our health at risk. People are sick and it is caused by breathing in toxic pollutants. How else would they be ill?"

On a cold night in December 2006, the Desert Rock issue came unexpectedly to a head when Elouise Brown of Doodá Desert Rock discovered a contractor affiliated with Sithe Global Power doing exploratory water drilling on grazing land permitted to Alice Gilmour, an elder in her eighties. Brown blocked the contractor's pickup, and Gilmour, members of Brown's extended family, and others joined her blockade. Brown appealed for help from other Navajo, and the blockade continued in subzero weather.

One visitor wrote: "They had a small, white tent that the grandmas were trying to stay in, but the wind blew through it; and they made a wood stove out of a 55-gallon drum, but the wind was blowing the smoke back into their tent; and the grandmas were having a hard time."

Soon videos of the blockade were being watched around the world, and supporters arrived to reinforce the protest. On December 22 police forcibly removed protesters from the road, but they established a nearby protest campsite and vigil that was still occupied nearly a year later.

At public hearings on the Draft Environmental Impact Statement in late July 2007 in several towns in Navajo territory, scores of local residents expressed vehement opposition to the plant. That month, Diné CARE sued the federal Office of Surface Mining for approving an expansion of the Navajo Mine to fuel the plant, and New Mexico governor Bill Richardson added his opposition to the plant. The next month, the Mountain Ute Tribal Council unanimously passed a resolution opposing construction, and in September the EPA expressed concerns about the thoroughness of the Bureau of Indian Affairs' draft environmental impact statement.

In September 2007 the construction contract was granted to the Fluor Corporation, but opponents continued to explore other avenues for slowing or blocking the project. One was to defuse the commonly cited argument by tribal officials that the plant would generate new jobs for an area with unemployment rates above 40 percent and poverty rates close to 50 percent. Was there a different course of economic development that would not exact such a terrible toll in illness and environmental degradation? To raise that option as a real possibility, Diné CARE presented Sithe with a report contrasting the development of the coal-fired plant with a clean energy scenario. The study based its argument on principles of Navajo ethics directing humans to live in harmony with the environment.

Meanwhile, a formal effort had been brewing that could provide the finances to underwrite such an alternative energy path. That effort grew out of the shutdown of the Mohave Generating Station in 2005 due to a Clean Air Act lawsuit and resolutions passed by both the Navajo and Hopi tribes ending Peabody's use of water from the Black Mesa aquifer to send the coal by slurry to Mohave. Because Mohave had been the highest emitter of sulfur dioxide in the western United States, shutting it down produced a windfall to the plant's owners in the form of pollution credits under the U.S. Acid Rain program administered by the Environmental Protection Agency. After the closure of Mohave, those credits began accumulating at the rate of an estimated $30 million annually. An alliance of groups calling itself the Just Transition Coalition (JTC) began working to secure the credits for tribal use by establishing a renewable energy infrastructure that would be partially owned by tribal communities and that would provide electricity, income, and jobs. The coalition included the Indigenous Environmental Network, Honor the Earth Foundation, Apollo Alliance, Black Mesa Water Coalition, To'Nizhoni Ani, Grand Canyon Trust, and Sierra Club.

The Just Transition Coalition proposed that annual revenues from the sale of pollution credits from the Mohave plant be reinvested in renewable energy on tribal lands, such as wind and solar plants, as well as be used to help offset the economic burden of lost coal royalties and jobs. In a formal motion to the California Public Utilities Commission, the coalition asked that the funds be allocated as follows: 30 percent for local villages and chapters to invest in solar, wind, and ecotourism; 10 percent for job retraining; 40 percent for alternative energy development and production; and 20 percent for tribal governments to help sustain programs cut due to loss of royalty income.

Throughout 2008 and into 2009, prospects for defeating Desert Rock appeared to steadily improve as project financier Blackstone suffered setbacks in the global financial crisis. In March 2009 Reuters reported that Blackstone CEO Steve Schwarzman had been forced to give himself a 99 percent pay cut as the private-equity firm posted a $1.33 billion loss. Estimated costs for Desert Rock had risen from $1.5 billion in 2003 to $4 billion in 2009, a sum that Blackstone now seemed less likely to be able to underwrite. Even as its stock market value plummeted from over $35 per share in early 2007 to less than $4 per share in early 2009, Blackstone was losing friends even more rapidly in the New Mexico state legislature, where an $85 million tax break for Desert Rock that had failed to secure passage in 2007 failed to win even a single sponsor in 2008. At both the New Mexico Environment Department and the EPA, regulators announced decisions to take a new look at the project's previously approved air permit because of ozone, carbon dioxide, and other issues.

Most ominously for the future of the project, potential buyers of Desert Rock's power were turning elsewhere for their future power needs. California utilities had already turned a cold shoulder to Desert Rock because of a new state law prohibiting the purchasing of power from coal plants that did not employ carbon capture and storage technology. Arizona Public Service, another potential buyer, had stated its intention to move away from coal toward solar energy. Only one real friend remained: the Navajo tribal government, which held out hope that the $50 million annual revenue stream promised by Schwarzman to the tribe could be revived. Ironically, money was now actually flowing from the tribe to the project, as the Navajo Nation racked up $110,000 in legal fees defending Desert Rock's permit applications. While Navajo Nation president Joe Shirley Jr. continued to win key Tribal Council votes securing right of way for the transmission lines required by the plants, closer votes on amendments to the transmission line legislation showed increasing uneasiness within the council.

As of mid-2009, Desert Rock still survived as a proposal, but activists opposing the project were hopeful. "We have drawn the line in the sand," said Dáilan Long.

Here's how a Navajo electrician summarized his feelings toward the project:

> "I worked on plenty of power plants in California and Arizona, but when one hit home I decided against it. Power plants are dirty, but the pay is good, yet I opposed Desert Rock. We are already dealing with two coal-fired power plants in the San Juan basin, and my mother, brother, and nieces are asthmatic and many people are sick with diabetes. I believe all this newly arrived disease comes from

breathing in chemicals from the power plants which slowly kills the inside organs. I just wish the Navajo Nation president and the council delegates could find something else in place of the power plants, coal mines, and oil fields. Our reservation is getting to be a dump yard for energy companies. We will be helping our president Joe Shirley and his council delegates digging graves for our future. We have to put a stop to this crazy genocide on Navajo land. We need help to put a stop to all this mess."

■

FOURTEEN

Cowboys Against Coal

■

ALTHOUGH OTHER REGIONS ARE more famously associated with coal, the largest coal reserves in the United States are actually located in the Northern Plains. If coal were gold, then the Gillette Field in Wyoming's Powder River Basin, sporting seams as thick as a hundred feet, would be Fort Knox. All ten of the largest coal mines in the United States are clustered here, just a few miles apart from each other. Two of these mines, the Black Thunder Mine and the North Antelope Rochelle Mine, together produce more coal than the entire state of West Virginia. Though not as fully exploited, Montana's coal reserves are even larger than Wyoming's, and according to some estimates they nearly equal the reserves of China.

Climate modelers knew that what happened in the Northern Plains had immense consequence for the future of life on Earth. If Wyoming and Montana, with more coal than all the states east of the Mississippi combined, were fully mined, the impact on atmospheric levels of carbon dioxide would be catastrophic.

The coal of the Northern Plains first placed the region in the crosshairs of national energy planners in 1971, when a report entitled the *North Central Power Study* sent shock waves throughout communities in the region. The study forecast a massive expansion in strip mining and electricity generation, including twenty-one new power plants in Montana alone. Facing the prospect of an industrial tsunami, a coalition of Montana ranchers and environmentalists organized the Northern Plains Resource Council (NPRC) in 1972. NPRC provided research and organizing resources for numerous county-level groups, each with its own autonomous operations.

Among the founding members of NPRC were ranchers who had survived everything the prairie had in its bag of delights—drought, dust storms, cattle-freezing blizzards—and weren't particularly intimidated by coal company flacks and lawyers. In his account of the coal fight, *The Rape of the Great Plains*, Montana historian K. Ross Toole described Bull Mountain rancher Boyd Charter:

> Boyd Charter (age sixty-six), who runs six hundred cows on fifteen sections of rangeland in the Bull Mountain area north of Billings, has a face that looks like the land he lives on. It is deeply lined and creased, the nose is large and a little bent, two lower teeth are missing, and the startlingly direct eyes are slightly hooded. Charter is clearly a man to be approached with caution, though ... he is a gentle man and a gentleman. One of the vice-presidents of Consolidation Coal Company did not approach him with caution. Charter recalls, "I told that son-of-a-bitch with a briefcase that I knew he represented one of the biggest coal companies and that he was backed by one of the richest industries in the world, but no matter how much money they came up with, they would always be $4.60 short of the price of my ranch."

The NPRC organizing model spread into the neighboring states of North Dakota and Wyoming, where the Dakota Resource Council and the Powder River Basin Resource Council formed on parallel lines. The three groups formed the Western Organization of Resource Councils, which eventually added four more statewide groups—Oregon Rural Action, Idaho Rural Council, Western Colorado Congress, and Dakota Rural Action—encompassing forty-five local groups and ten thousand members. Over time, this far-flung coalition developed into one of the most effective grassroots environmental networks in the United States.

The first big power plant fight in the Northern Plains involved Montana Power Company's Colstrip complex in southeastern Montana. The ranchers and their allies, including the Northern Cheyenne tribe, lost the battle, and by 1986 Colstrip had expanded to four generating units. But the Colstrip fight spawned a wider movement in Montana that secured the passage of some of the strongest state-level environmental legislation in the country. Above all, Montanans were determined to prevent the coal industry from dominating the state the way that the hard-rock mining industry, especially the Anaconda Copper Mining Company, had done for much of the twentieth century.

Despite Montana's vast reserves, the state's coal mines were producing less than a tenth the tonnage of neighboring Wyoming at the time the Bush administration began its push for a new wave of coal plants. One reason had to do with transportation infrastructure. An immense rail complex had been constructed to bring coal from Wyoming's Powder River Basin to power plants as distant as Florida, but fewer rail lines extended into Montana's coalfields. Generally, the development of infrastructure—not just rail lines but also high-voltage transmission lines, water pipelines, and pipelines for transporting synthetic gas or liquids—is an incremental process. Each mine or plant built in a coal region makes the next facility easier to site, and so forth. By following the path of least resistance and building mine after mine close to existing rail lines in Wyoming, the mining industry had failed to create the necessary transportation foothold in Montana. That gave anti-coal activists even more of an incentive for keeping any new development out of the state.

During 2007 and 2008, the proposal that figured most prominently in Montana politics was the Highwood power plant, proposed by the Southern Montana Generating & Transmission Cooperative (SMGTC), which was slated to sell part of its output to several Montana cities. Grassroots opposition to the Highwood proposal arose quickly, and the intensity of the response caused cities of Helena and Missoula to back away from the project. The erosion of support did not stop Highwood, but it was a first step in undermining it. Since the project had already secured most of its necessary environmental permits, the best hope for opponents was to focus on the project's financing. The Rural Utilities Service (RUS), historically a strong supporter of coal plants, backed loans for Highwood.

Along with several allies, the Montana Environmental Information Center (MEIC) sued to stop the RUS from lending money to Highwood on the basis that the federal loan program had not been subjected to a formal environmental review. Surprisingly, the RUS caved quickly to the pressure, announcing in February 2008 that it was placing loans to *all* coal plants on hold. Officials cited the "inherent risks associated with compounded delays" and concerns about financial feasibility in light of increasing cost estimates.

Meanwhile, opponents appealed Highwood's air permit to the Montana Board of Environmental Review (BER), raising health concerns and calling for further study of particulate emissions. In a 6–1 ruling in April 2008, the BER ordered more research on particulates smaller than 2.5 microns in diameter, known as PM2.5. The ruling made the board the first regulatory body in the nation to order separate measurements and emissions controls for PM2.5.

SMGTC continued to pursue the project, but support was clearly eroding. The final decision would not come for another full year—January 2009—when the SMGTC announced that it was canceling the plant and instead building wind power with natural gas backup.

Another project, a coal-to-liquids plant proposed for siting at Malmstrom Air Force Base near Great Falls, posed a different set of concerns. The project had the enthusiastic support of Montana governor Brian Schweitzer, an up-and-coming star within the Democratic Party and a synfuels booster. Since Malmstrom was a military project, opponents feared that it could potentially be exempted from environmental regulations.

As is typical in such military-industry projects, the advocates for the Malmstrom project seemed one minute to be working at Pentagon desks and the next minute at private contractors that would benefit if the project were built. One such revolving-door operative was Ron Sega, the Air Force undersecretary who flew the first Air Force jet powered by synfuels in September 2006. In December 2007 disclosure forms revealed that Sega had left the Air Force and joined the board of synfuels technology developer Rentech.

Much was at stake. The U.S. Air Force uses more than half of the fuel consumed by the U.S. government. In 2007 the Air Force spent $5.8 billion to buy 2.6 billion gallons of fuel. For every $10 increase in the price of a barrel of oil, the amount the Air Force spends on fuel rises by $600 million. Part of the concept being promoted by Ron Sega and others was for the Air Force to certify its fleet of nearly six thousand aircraft to use a 50:50

blend of synthetic fuel and petroleum-based jet fuel by 2011. If such plans became a reality, companies like Rentech would hit the jackpot. To build support, Rentech hired the lobbying firm of Brownstein Hyatt Farber Shreck to work the Hill.

The scale of the proposed Malmstrom plant was immense. Each day 20,000 tons of coal and 10 million gallons of water would enter the plant, and 20,000–30,000 barrels of fuel, 1200–2400 megawatt-hours of electricity, and 15,000 tons of carbon dioxide would exit it. Developers promised that the carbon dioxide would be pumped into deep underground formations, but details were not forthcoming.

On January 30, 2008, Congressman Henry Waxman, chairman of the House Committee on Oversight and Government Reform, and Tom Davis, ranking minority member of the committee, wrote to Defense Secretary Robert Gates, requesting information on how the Department of Defense's plans for coal-based synfuels would comply with new greenhouse gas limits imposed on federal agencies by the Energy Independence and Security Act of 2007 (EISA). According to Section 526 of the law:

> No Federal agency shall enter into a contract for procurement of an alternative or synthetic fuel, including a fuel produced from nonconventional petroleum sources, for any mobility-related use, other than for research or testing, unless the contract specifies that the lifecycle greenhouse gas emissions associated with the production and combustion of the fuel supplied under the contract must, on an ongoing basis, be less than or equal to such emissions from the equivalent conventional fuel produced from conventional petroleum sources.

Despite the apparent restrictions contained in EISA, opponents of the Malmstrom plant found little reassurance. Considering the powerful support enjoyed by the project, Section 526 might prove to be little more than a speed bump. Fortunately, a combination of unexpected circumstances arose that derailed the Malmstrom proposal. As a worldwide recession derailed economies around the world, oil prices plummeted and coal-to-synfuel projects became increasingly shaky. The election of President Obama also promised to at least somewhat curb the enthusiasm for coal that had characterized the Bush presidency. On January 29, 2009, with little fanfare, Air Force officials announced that they would no longer pursue development of the Malmstrom project. The explanation was quirky. Had the plant been built, its tall structures would have created helicopter-flight safety issues. In addition, operation of the plant would potentially have "created conflicts with the missile wing's mission, including reducing security near the nuclear weapons storage area and an 'explosive safety arc' surrounding it, and interfering with missile transportation operations on internal Malmstrom roads."

In the wake of the stroke of fortune, opponents such as Anne Hedges of the Montana Environmental Information Center breathed a sigh of relief. The fact that the project had gotten as far as it did was a reminder of what a tempting opportunity Montana continued to present to energy developers. The fact that developers had tripped over their own logistics was a reminder that sometimes the fates do smile on Mother Earth.

East of Montana, energy companies were targeting the Dakotas for new coal development. North Dakota's coal, though abundant, is a low-grade variety known as lignite that is too poor in quality to ship long distances. As a result, mining in North Dakota had clustered in a strip alongside the Missouri River, where cooling water was available for a half-dozen power plants, built from the 1950s to the early 1980s. Every year, the mines associated with those plants consumed thousands of acres of valuable cropland, and power plants emitted plumes containing sulfur dioxide, mercury, and other toxins that drifted eastward across Minnesota, Wisconsin, and the Great Lakes. The study *Dirty Kilowatts* had listed five of the central North Dakota coal plants among the fifty worst emitters of carbon dioxide and mercury in the country.

Mining and energy companies had long wanted to expand the area of concentrated coal development farther into southwestern North Dakota, but each such proposal sparked resistance. A chokepoint for industry development was the Theodore Roosevelt National Memorial Park, whose Class I air quality status prevented plants from being sited nearby. North Dakota's pro-coal state government sought to replace federal air modeling methods with new models that would allow more plants to be built in areas with sensitive air quality. In response, the Dakota Resource Council sued twice to block the weaker standards. DRC lost both cases, and in the final months of the Bush administration the EPA announced plans to approve North Dakota's weaker air models for use across the country. But less than a month before Bush left office, the EPA admitted that it had run out of time to weaken the air standards. Terrence Kardong, a Benedictine monk who had worked on coal issues for three decades, declared "a win for the mouse" and provided a pithy summary of the long fight: "The Bush gang finally gave up and we did not."

Of all the power plant fights in the region, the most intense was the struggle over the Big Stone II plant proposed for South Dakota near the Minnesota border. Initially, the plant was sponsored by seven utilities, including lead developer Otter Tail Power, Central Minnesota Municipal Power Agency, Great River Energy, Heartland Consumers Power District, Missouri River Energy Services, Montana-Dakota Utilities Co., and Southern

Minnesota Municipal Power Agency. Facing off against the utilities was an even larger coalition of citizen groups, including Beyond Big Stone II, Dakota Resource Council, South Dakota Clean Water Action, Sierra Club Northstar Chapter, Minnesota Center for Environmental Advocacy, Union of Concerned Scientists, Izaak Walton League, Land Stewardship Project, Wind on the Wires, Fresh Energy, and Clean Up Our River Environment (CURE).

Opponents of Big Stone II pursued every available route to voice their protests. CURE built a miniature coal plant, propped it between two canoes, and entered the float (along with a ranting coal baron) in the annual River Blast Flotilla on the Minnesota River. High school students descended on the state capitol in St. Paul, where they lobbied legislators and grilled Governor Tim Pawlenty's political deputies. Eight Minnesota legislators wrote to Microsoft's Bill Gates, whose investment company owned a 9 percent stake in Big Stone II sponsor Otter Tail, inviting Gates for a visit to review renewable investment opportunities in Minnesota that would "align the values of your foundation with your investment strategy." James Hansen wrote personally to the governor, expressing opposition to the plant. Videos of children protesting the mercury emissions from the plant circulated on YouTube.

David Schlissel, an analyst at Synapse Energy Associates in Boston, developed one of the most persuasive arguments against Big Stone II. Schlissel noted that the utilities proposing the plant had failed to account for two types of risk. First, by failing to account for the likelihood that some kind of carbon-pricing legislation was likely to be enacted in the coming years, the sponsors had underestimated the cost of coal. In comparison, the cost of power from wind generators was highly predictable, since after the initial capital costs and ongoing maintenance costs, wind generators did not require any sort of fuel supply. Second, the sponsors had overestimated the reliability of the Big Stone II plant because they had failed to recognize a growing number of transportation and other bottlenecks that had already caused periodic interference with supplies of coal coming from the Powder River Basin. For example, in 2005 two train derailments produced a domino effect of coal shortages at power plants located far from Wyoming, causing $2 billion in losses.

Schlissel's twin arguments went to the heart of the supposition that burning coal is the cheapest, most reliable way of generating power. Over time, opponents of coal plants elsewhere would further develop those arguments. Working from her home office in Boulder, Colorado, Leslie Glustrom, a member of the No New Coal Plants list, delved deeply into studying the topic and came to the conclusion that utility planners and lobbyists had been painting far too rosy a picture of future coal availability. Contrary to the common assumption that the United States has a 250-year supply of coal, Glustrom found analyses by the U.S. Geological Survey (USGS) pegging the supply at actual operating mines at approximately nineteen years. That number, of course, could be increased if new mines were to open. But doing so at the rate needed to supply currently operating plants would not be easy. East of the Mississippi, most states had been experiencing a long-term decline in production levels as the easiest coal seams were mined out. As regulators increased their scrutiny of mountaintop removal mining, eastern production would continue to fall.

West of the Mississippi, reserves were more abundant, but obstacles existed to expanding current mines. For example,

USGS review of the Gillette Coal Field in the Powder River Basin, the source of 40 percent of the nation's coal, reduced the estimated reserve at current prices to a mere 10 billion tons, down from an estimate of 23 billion tons in 2002. Key to the reduction in coal reserves was the recognition that the vast majority of coal in the Powder River Basin either was buried too deep to be economically recovered or was unavailable for other reasons, such as conflicts with roads, towns, or environmentally sensitive areas.

As opposition to Big Stone II multiplied, two of the cosponsors of the project, Central Minnesota Municipal Power Agency and Great River Energy, got cold feet, exiting the project in the fall of 2007. That left the plant undersubscribed by about 27 percent and meant that it would need to be downgraded in size. Meanwhile, projected construction costs were continuing to increase.

On May 9, 2008, two administrative law judges recommended to the Minnesota Public Utilities Commission that the transmission line permit for the plant through western Minnesota be denied, based on their conclusion that conservation and load management measures could more economically satisfy the demand for electricity. The decision came as a further blow to the project. In October another blow arrived when the Minnesota Public Utilities Commission received a report from Boston Pacific Co. of Washington, D.C., saying that the utilities had underestimated construction costs and overestimated the costs of alternative energy sources.

On January 23, 2009, three days after the Obama administration took office, the EPA filed objections to South Dakota's air permit for Big Stone II. In April, South Dakota went ahead and issued the air permit, but the project continued to face

6

numerous other permitting hurdles and legal challenges. Meanwhile, support for the plant eroded further in July 2009, when the municipal utility for Elk River, Minnesota, backed out of the project. In September came bigger news: Otter Tail Power, the main sponsor of the project, also backed out, opting to focus instead on developing cleaner alternatives. Already, Otter Tail had committed to developing 180 megawatts of wind power, making it, relative to its size, one of the most wind-reliant utilities in the country. With only four participating utilities left, the odds that the plant would be built were becoming increasingly slim.

■

FIFTEEN

Sierra

■

IN MICHAEL ONDAATJE'S NOVEL *The English Patient*, a character named Kip works for the British army as a sapper, performing the work of defusing unexploded bombs and land mines. The job calls for patience, nerves, and a bit of luck. If coal plants are planetary time bombs, then by all accounts the best sapper in the movement to defuse them was a Sierra Club employee named Bruce Nilles, who initiated and led the club's national coal campaign out of a small office off Federal Street in Madison, Wisconsin.

While still an undergraduate at the University of Wisconsin, Nilles had written a paper about how to clean up the local Charter Street coal plant. Nobody took the idea seriously, but even after leaving the state, the Charter Street plant remained in the back of his mind. In 2002 Nilles signed on with the Sierra Club and became the sole staffer for the club's coal campaign, focusing initially on Illinois.

Although it was not apparent at the time, this one-state effort, which eventually became a nationwide project, marked a milestone in the history of the Sierra Club. For most of the

116 years of its existence, the club had been a fairly genteel organization of nature lovers that concentrated its energies on saving wild, unspoiled places like the Grand Canyon or California's Hetch Hetchy Valley. That changed with the arrival of firebrand David Brower as president and with the concurrent expansion of the environmental movement in the 1960s and 1970s, during which the club's focus widened accordingly to include pushing for federal legislation that benefited not just remote and scenic areas but the environment as a whole. But even Brower had never taken on a challenge as big as the all-out mobilization to stop 151 coal plants across the United States. Considering the potential of those plants to push the global climate into dangerous warming, the stakes of the campaign were no less than the fate of the planet itself.

Among the major groups that make up Big Green, the Sierra Club was the only organization to develop a strategy on coal that fully matched the level of alarm being sounded by climate scientists. By 2007 James Hansen and his colleagues were insisting that coal needed to be completely phased out by 2030. Accomplishing that objective didn't just mean encouraging efficiency, pushing for clean power, or raising the cost of fossil energy, though all of those were important measures. It meant blocking each new coal-fired power plant from going beyond the drawing boards, then moving on toward phasing out every existing coal plant. An obsessive focus on coal plants was what Bruce Nilles brought to the Sierra Club, and it was what set the Sierra Club apart. At any given time from 2007 onward, the club was directly involved in regulatory interventions or lawsuits affecting dozens of proposed projects across the country.

The industrial heartland was an appropriate place to start. The region, including Ohio, Illinois, Indiana, Missouri, Pennsylvania, and Michigan, has the most intensive coal generation in the country. No state burns more coal than Ohio. Indiana comes in second, Illinois fifth. Erik Shuster's list of 151 proposed new coal plants included sixteen plants in Illinois alone (largely due to misguided efforts by the state legislature to use coal as a means of economic development), and twenty-one more plants in Ohio, Indiana, Michigan, Missouri, and Pennsylvania.

Even if no new coal plants were built, citizens in the Midwest already pay a heavy price in health effects for the region's reliance on coal. Of the twenty-four thousand people estimated to die prematurely in the United States due to fine particles from power plants, a third are in the six industrial heartland states.

In Illinois, the first target of Sierra's coal work, the club eventually was able to claim victory against all but five of the sixteen proposed plants. It was a remarkable accomplishment, considering that the state's political establishment, led by Governor Rod Blagojevich, was solidly behind coal. In 2002 Illinois had created a Coal Revival Program to support new coal-fired plants. In July 2003 the state expanded its support for coal with $300 million in state-backed bonds to help finance the construction of "advanced technology" coal-fueled projects.

One of the toughest coal plant fights was in Franklin County, where EnviroPower had already started construction of a 600-megawatt plant when a U.S. District Court judge ruled in favor of Sierra's objections to the air permit. To make its case at the appellate level, EnviroPower hired Harvard celebrity lawyer Alan Dershowitz, who accused Sierra of undermining national security. "The Sierra Club's latest salvo to stop all coal-fired power plants in the Midwest threatens America's energy independence," said Dershowitz.

Despite the histrionics of Dershowitz, Sierra prevailed in the U.S. Seventh Circuit Court of Appeals in Chicago, and EnviroPower was forced to abandon the facility. Even more controversial was the Taylorville Energy Center, which had managed to gain the support of several environmental groups because it planned to use the new integrated gasification combined cycle (IGCC) technology. The groups that supported the project, including the Citizens Utility Board, the American Lung Association, and the Illinois Clean Air Task Force, believed that implementing IGCC technology at a commercial scale was a step toward the Holy Grail of climate-friendly coal usage. Sierra and most other groups opposed the project, since the developers had no plans to actually capture and store the carbon dioxide emissions.

Even as Sierra's coal work expanded beyond the Midwest into Kansas, Florida, Nevada, and other states, Bruce Nilles continued pursuing his long-standing goal of closing Madison's three old coal plants. In 2005 Sierra had kicked off a campaign to shut down the largest and dirtiest of the trio, Madison Gas & Electric's Blount Street facility, beginning with a citywide educational campaign. More than two hundred people showed up at a City Council meeting demanding closure. Soon, Madison Gas & Electric announced that it would cease burning coal at that site in 2010, and Sierra shifted its work to the second-biggest source of pollution, the Charter Street plant.

For two years, Nilles and others negotiated with officials from the University of Wisconsin, but when they realized that the talks were failing to make progress Sierra began investigating the plant's compliance with air-quality regulations. Finding multiple violations, Sierra sent the state of Wisconsin a letter informing it of the problems in November 2006. After another round of stalling by the utility, Sierra sued both the university and the state in federal court. The suit revealed that the operators had significant compliance problems statewide and no oversight by the Department of Natural Resources. When U.S. District Court Judge John Shabaz ruled in favor of Sierra, the state finally agreed in late 2007 to reduce coal use at the Charter Street plant.

During 2007 and 2008, the number of proposed coal plants in which the Sierra Club was involved multiplied across the country. As described in chapter 7, the fight over the expansion of the Holcomb coal plant in Kansas was the most significant of these, but at the same time the Sierra Club, together with local allies, was involved in regulatory proceedings or litigation against plant proposals in Texas, Missouri, Florida, Kentucky, Arizona, Oklahoma, Iowa, Washington, Utah, Georgia, Illinois, Wisconsin, Montana, Nevada, Michigan, Louisiana, Arizona, and other states.

In 2008 Sierra upped the ante from opposing individual coal plants to organizing against an entire company, Houston-based Dynegy Inc. Across the United States, Dynegy, together with its joint venture partner LS Power, was planning more new coal plants than any agency or utility. If built, the plants would add 44 million tons of carbon dioxide to the atmosphere each year. Branding Dynegy "America's Coal-Fired Polluter Number 1," Sierra kicked off its campaign in late February with mass call-ins to Dynegy headquarters originating from twenty states. In May, a hundred Sierra activists showed up at Dynegy's annual meeting and delivered 10,000 letters and emails to the company's CEO, Bruce Williamson, urging the company to redirect its investments toward cleaner sources of energy. The campaign quickly hit a nerve with Williamson, who complained that Dynegy was being unfairly picked on. It probably didn't help Williamson's morale that he had also just been picked as one of five executives to receive Fossil Fool of the Year awards.

Dynegy was a ten-year-old power company that had already had one near-death experience, getting caught up in charges of price fixing and other fraudulent practices during the California electricity crisis of 2000 and again in the wake of the Enron debacle in 2002. Williamson, who came on board to replace founder Charles Watson, was credited with saving the foundering company by exiting natural gas and moving into coal. That move placed the company on a collision course not only with Sierra but with local groups in over a half-dozen states. Before the launching of the Sierra Club campaign, five Dynegy projects had already bit the dust in Illinois, Oklahoma, Virginia, South Carolina, and New Jersey. The remaining six were the Longleaf plant in Georgia, the Power Elk Run plant in Iowa, the Midland plant in Michigan, the Plum Point plant in Arkansas, the Sandy Creek plant in Texas, and the White Pine plant in Nevada.

Meanwhile, Dynegy had also come under pressure as one of five energy companies subpoenaed in 2007 by New York attorney general Andrew Cuomo under New York State's Martin Act, a 1921 securities law that gives the state broad access to corporate financial records. The purpose of Cuomo's investigation was to determine whether Dynegy and the other companies were adequately informing investors about the financial risks connected to their emissions of global warming gases. In October 2008 Dynegy agreed to disclose information about how global warming might affect its business practices, including explaining the potential consequences to investors if federal rules are adopted to limit carbon dioxide emissions. Dynegy also agreed to report its efforts to mitigate carbon dioxide emissions, estimates of its financial liability in settling possible lawsuits related to climate change, and the potential impact of climate shifts on its ability to generate electricity.

On top of the pressure from activists and regulators came the worldwide financial crisis and economic recession, which looked likely to suppress demand for new power plants. During Dynegy's regular report to financial analysts in November 2008, Williamson admitted that "very little new power plant development is going on in the country and very little can be economically justified in the current environment." He hinted that economic conditions would likely slow the demand for power in the short term.

The following month, Williamson announced a major re-evaluation in Dynegy's coal plans. Citing the combination of economic problems and hardening opposition, the CEO said that Dynegy had decided to reassess its involvement in all six of the projects targeted by Sierra and other groups. In January 2009 Dynegy announced it was dissolving its development venture with co-developer LS Power. The official position of LS Power was that it intended to continue developing the projects, but opponents noted that the company, which had never built a coal plant, was less likely than Dynegy to move the projects forward. Sure enough, LS Power began canceling projects: Iowa's Elk Run plant in January 2009, the White Pine project in Nevada in March 2009, and the Midland plant in Michigan in April 2009.

While Sierra's organizers and some of its attorneys worked on the Stop Dynegy campaign, another legal team pursued a petition with broad potential implications aimed at forcing regulators to include the Supreme Court's ruling in *Massachusetts v. EPA* in all power plant siting decisions. The focus of the petition was the proposed Deseret Plant in Utah, which was subject to federal jurisdiction due to its siting on the Uintah and Ouray Indian reservation. Sierra attorneys David Bookbinder and Joanne Spalding argued before the EPA's Environmental Appeals Board that *Massachusetts v. EPA* required new plants like Deseret to use the "best available control technology" to minimize carbon dioxide emissions. In a ruling that sent shock waves through the utility industry, the appeals board agreed with Sierra, a decision that temporarily froze coal plant applications across the country while EPA developed regulations for implementing carbon dioxide controls.

In the spring of 2009, Bruce Nilles picked up stakes and moved the office of the Sierra Club's coal campaign to Washington, D.C. Having been voted Hero of the Year by the readers of the online magazine *Grist* and one of a hundred "Agents of Change" by *Rolling Stone*, Nilles had become something of a celebrity within the environmental movement. With the move to Washington and the arrival of the Obama administration, the focus of Sierra's coal work was now shifting toward national regulation and legislation. Nilles believed that the most promising avenue for action derived from the Supreme Court's decision directing the Environmental Protection Agency to begin regulating greenhouse gases. While the Bush administration had dragged its feet on implementing the decision, under Obama's EPA secretary Lisa Jackson the agency moved quickly to make a determination that carbon dioxide and other greenhouses gases "may be reasonably anticipated to endanger public health and welfare." That determination was crucial to taking further steps toward limiting carbon dioxide from coal plants, and Sierra mobilized its members to push the EPA toward an aggressive implementation of the finding, packing public hearings in Arlington, Virginia, and Seattle, Washington.

In Seattle a boisterous crowd of about two thousand people turned out for the event. Banners hung from the roof of the convention center and hundreds of people rallied outside, while inside an array of speakers turned a dry hearing into something closer to a celebration. Side by side with lawyers and technical experts who provided detailed analyses of the Clean Air Act were moms who brought children to the podium as "visual aids," students who testified wearing garbage bags, and activists from Appalachia who vividly described growing up near streams that ran black with coal dust. The hearing showed Sierra at its best—an organization capable of merging the talents of legal and scientific "paper jammers" with the efforts of passionate volunteer activists.

The club had much to cheer, having played a central role in a movement that had accomplished the seemingly impossible task of blocking coal plants by the score. The question, of course, was whether the movement could maintain that momentum, or whether the coal industry would find a way to regroup. One thing was clear: if the battle was shifting to Washington, the coal industry, with its immense lobbying resources and strong political connections, would have a home field advantage against Sierra and groups that drew their strength primarily from the grassroots. Under the Bush administration, there had been every reason for the grassroots movement against coal to work primarily at the local and state levels. Now Sierra and other groups had to move to the next level without losing their focus and becoming isolated from their base.

■

Taking It to the Streets

∎

IN THE WANING DAYS of 2008, the fight over coal seemed to briefly enter a strange zone of disconnection from reality— a surreal moment when debate over coal devolved into an argument over Christmas itself. At the Web site of the main pro-coal lobbying group, the American Coalition for Clean Coal Electricity (ACCCE), visitors were treated to the bizarre sight of lumps of coal, dressed as carolers, singing traditional Christmas tunes with lyrics such as:

> Frosty the coal man is a jolly happy soul...
> There must be magic in clean coal technology
> For when they looked for pollutants
> There was nearly none to see!

Climate blogger Joe Romm marveled at the inanity of the carols: "In the twisted minds of the industry Mad Men who put this together, it makes perfect sense to turn songs about the birth of Jesus into songs about clean coal.... I'd say clean coal had jumped the shark, but I think you have to actually exist first before you can become self-parody."

Joe Lucas of ACCCE responded, "I'll put my years as a Sunday school teacher, church deacon, and church musician up against just about anybody else when it comes to understanding hymnology and respect for religious traditions."

What happened next amounted to a jolt of reality—delivered in the middle of a cold and moonless night to neighbors of the 55-year-old Kingston Fossil Plant near Harriman, Tennessee. Less than an hour after midnight on December 22, 2008, Chris Copeland, who lived outside Harriman with his wife and children on Watts Bar Lake, was awakened by a noise that he described as "crashing and popping." Looking through his bedroom window, Copeland saw "waves of water going through the cove back here … debris, trees flowing through the backyard."

Not far from the Copelands' house, a 60-foot-high impoundment containing fly ash from the plant had breached its containment dike and flowed out onto three hundred acres of residential land. The fly ash clumped in soot-gray icebergs that floated across roads and nestled against backyard swing sets. As residents became aware that the fly ash was laced with mercury, lead, cadmium, beryllium, and a host of other toxins, the Tennessee Valley Authority, which operated the plant and the impoundment, faced angry questions.

It was King Coal's worst nightmare come true: toxic waste from coal flowing straight into the two-car garages of suburban Americans. At over a billion gallons, the spill was larger than any previous coal-related spill in American history. It was one hundred times the reported size of the Exxon Valdez disaster. Miraculously, there had been no loss of life.

For several days, the major media failed to grasp that something significant had happened. Matt Landon and other members of United Mountain Defense arrived on the scene and began organizing an action plan as well as contacting writers like Jeff

Biggers, who immediately began blogging about the disaster. On the Web, Twitter was abuzz with posts on the spill. Amy Gahran, a Colorado-based media consultant, threw herself into spearheading a national effort to make information on the spill available to other Twitterers. RoaneViews.com, a Web site for the community near the Kingston plant, participated in the Twitter campaign, as did the *Knoxville News-Sentinel* and the *Nashville Tennessean*. EPA staffer Jeffrey Levy provided agency maps and statistics on the plant. Then the mainstream media woke up, and photographs of the Tennessee sludge spill finally became a high-profile news item worldwide.

Barely a week after the accident, a Google search for the phrase "Tennessee spill" produced 2,280,000 results, making it one of the most prominently reported environmental catastrophes in decades. Press attention focused on the discrepancy between the industry's claims about coal being clean and the ugly reality on display in Harriman, Tennessee. Clearly, the "I believe" clean coal ad campaign had backfired. After Tennessee, the ads continued to run, but for many people those ads merely served as reminders of coal's actual impacts. Activists who had spent years struggling for some sort of coal waste regulation—none yet existed—saw that a window for legislation was suddenly wide open.

Meanwhile, the election of Barack Obama had raised hopes among environmentalists, and within weeks of taking the oath of office, Obama lifted expectations further as he signaled the intention of making a dramatic break from the Bush administration on the issue of climate change. Most of Obama's appointees looked promising, especially Stephen Chu, a California Nobel laureate who had described coal as "my worst nightmare," as secretary of energy. That impression was reinforced by a stream of positive steps by regulatory agencies in the weeks that followed the arrival of the Obamistas. These included the EPA's challenge to the air permit for Big Stone II, the U.S. Air Force's cancellation of the Malmstrom Air Force Base coal-to-liquids project, and the EPA's initial steps toward regulating carbon dioxide and five other global warming gases under the Clean Air Act. More moves to regulate coal would be announced during the remainder of the spring, including a wide review of over two hundred mountaintop removal mining permits and initial steps toward regulating fly ash.

Anticipating the shift toward Washington, D.C., the Reality Coalition, which included the Sierra Club, the National Wildlife Federation, the Natural Resources Defense Council, the League of Conservation Voters, and Al Gore's Alliance for Climate Protection, blanketed the capital with an advertising campaign that ridiculed the idea of clean coal. The Reality Campaign countered the coal industry's simplistic clean coal message with an equally simple response: clean coal doesn't exist.

In one ad, created by the Crispin Porter & Bogusky agency, a man wearing a hard hat and holding a clipboard invites the viewer to take a tour of a clean coal facility. Opening a door, he steps onto a barren desert. "The machinery is kind of loud," he shouts above the wind, "but that is the sound of clean coal technology."

The ad ends with the words: "In reality, there is no such thing as clean coal in America today."

The introduction of the Reality Campaign, which began shortly before the Tennessee sludge disaster, could not have been better timed. The targeting was also good: heavy expenditures on billboards throughout Washington, D.C., ensured that federal policymakers charting strategy for the incoming Obama administration would get the message. Anti-coal activists took heart: after years of deciding whether to fight the coal industry or make a deal with it, perhaps the environmental movement was beginning to make up its collective mind that fighting, rather than compromising, was the best strategy.

At the end of February, I flew to Washington, D.C., to participate in the first nationwide protest aimed at coal: a nonviolent blockade of a 99-year-old coal plant that operated just blocks from the Capitol itself. It promised to be a watershed event, for various reasons. One was that it marked a shift from protesting new coal plants to phasing out existing ones. Having succeeded in sidetracking most of the 151 new coal plants that had been proposed less than two years earlier, the next challenge was to phase out the existing fleet of coal plants. On the CoalSwarm wiki, which had grown to include 1,500 articles and had attracted over a million visits, reader statistics showed that the movement was pivoting rapidly toward assessing this new challenge. Whereas earlier the most popular pages on the wiki had been "Coal plants canceled in 2007," now people visiting the site were most likely to read the page entitled "Existing U.S. coal plants."

If phasing out existing coal plants was the goal, the Capitol Power Plant was a good place to start. Among a fleet that consisted of six hundred aging plants, half of which were built before 1965, the Capitol Power Plant represented the oldest of the old, having been commissioned by an act of Congress in 1904 and completed in 1910.

But old didn't necessarily mean feeble. Though the Capitol Power Plant hadn't produced a watt of electricity since 1952—just steam and refrigeration for the Capitol Complex—the old horse was still delivering the same chest-crunching, asthma-inducing kick, literally killing people in surrounding neighborhoods.

It was a good example of the fact that replacing dirty power plants with clean alternatives wasn't just a crucial step toward solving the climate crisis, it also had major public health benefits. A 2004 study by the Clean Air Task Force estimated that 515 people were dying annually in the D.C. metropolitan area because of power plant emissions, fifth among all U.S. cities. In 2002 the Capitol Power Plant was responsible for 65 percent of the PM2.5 particulate pollution produced by all point sources in the District of Columbia. The consequent toll of premature deaths was falling disproportionately on a low-income, largely African American population. Hill Residents for Steam Plant Conversion, a neighborhood group, had so far been unsuccessful in getting the plant shut down, but there had been some progress in Congress, where Nancy Pelosi had initiated a program to replace coal with natural gas for the portion of the Capitol Power Plant that serves the House of Representatives.

On the Senate side, things didn't look as promising. Since 2000 two Senate leaders, Mitch McConnell (R-KY) and Robert Byrd (D-WV), with a combined tenure of seventy-three years, had blocked the Senate from eliminating coal at the plant. In May 2007 CNN reported that International Resources Inc. and the Kanawha Eagle mine had received contracts to supply a combined 40,000 tons of coal to the plant over the next two years. The two companies had given $26,300 to the McConnell and Byrd campaigns during the 2006 election cycle.

Considering the death toll from air pollution, the destructive mining, the dirty money, and the climate impacts, the Capitol Power Plant was a pretty good microcosm of what was wrong with U.S. coal policy.

In setting a date for the blockade against the Capitol Power Plant, the organizers chose the weekend coinciding with

PowerShift, a huge youth climate conference that was expected to bring over twelve thousand young activists to D.C. Gearing up for the March 6 action, organizers from Rainforest Action Network, Greenpeace, Chesapeake Climate Action, and other groups worked to spread the word about the blockade and to recruit, organize, and train activists in the principles and techniques of nonviolent civil disobedience.

Matt Leonard, one of the organizers, wrote:

> We aim to create an action framework that is accessible to all—from students, to elderly, to parents, to notable public figures and beyond. We envision a structured event that includes agreed-upon action guidelines, extensive training on non-violence, and a respectful tone that participants would be asked to abide by. We will have a legal team organized to support participants and will have prior discussion with authorities as to our non-violent intentions.

Leonard and the other organizers knew that the decision to conduct civil disobedience, however nonviolent the intent, meant walking a tightrope. While any such action cannot be carried out without a certain degree of intensity, at the same time the message needed to be broad enough to attract a spectrum of groups. The tone needed to be militant but tightly disciplined, since even a single act of violence could undermine the entire project.

Within the climate movement, the wisdom of such militancy was far from universally accepted. As the organizers of the Capitol Power Plant action approached individual groups for endorsements, leaders were forced to take sides on the question of whether they would publicly support an action that would openly involve civil disobedience.

Some felt that even a slight possibility of violent disruption made the action ill advised. To others, the timing was wrong. They believe that mounting a civil disobedience action in

Washington within the first one hundred days after the inauguration would alienate the new administration by appearing to be a punishment aimed at Obama before he had even had time to flesh out his policies toward coal. Supporters of the action countered that unless the movement moved quickly to assert its strength, the Obama administration would fall prey to the same utility and coal interests that had long controlled the back rooms of regulation, legislation, and policy.

There was a hint of generational tension in the response to the action. Those declining to participate (including all of the major environmental groups) as well as those claiming that the action was too early, too risky, or too militant tended to be middle-aged "establishment" environmentalists. Most of the organizers (with the notable exception of Ted Glick, a veteran of the Vietnam-era peace movement) were younger than thirty. To them, the action was infused with a sense that the time had come to make a clear break from the ineffectual "insider" tactics of the past. Instead, an "inside/outside" approach was needed to allow more forceful pressure to be applied to the political process. The organizers wrote:

> We can determine the fate of our generation. We know there is a climate crisis and we know we have to stop it. We've organized, we've lobbied, we've passed policies, we've educated, we've agitated, and still our government has not recognized the scope and urgency of global warming. We know we have the capacity to transform our society. What we lack is the political will.

> But now there is a new administration and a new Congress, which gives us another chance. We have a window, but we must open it. Together.

> Like the movements that have come before us, we have an opportunity to send a powerful message of urgency through peaceful civil disobedience. There has never been an American tradition more noble, and it is needed now more than ever.

But while organizing the action was mainly in the hands of young activists, a number of senior figures in the environmental movement lent their support. Two writers, Bill McKibben and Wendell Berry, penned the following call to action:

> Dear Friends,
>
> There are moments in a nation's—and a planet's—history when it may be necessary for some to break the law in order to bear witness to an evil, bring it to wider attention, and push for its correction. We think such a time has arrived, and we are writing to say that we hope some of you will join us in Washington D.C. on Monday March 2 in order to take part in a civil act of civil disobedience outside a coal-fired power plant near Capitol Hill... The industry claim that there is something called "clean coal" is, put simply, a lie. But it's a lie told with tens of millions of dollars, which we do not have. We have our bodies, and we are willing to use them to make our point... It's time to make clear that we can't safely run this planet on coal at all... This will be, to the extent it depends on us, an entirely peaceful demonstration, carried out in a spirit of hope and not rancor. We will be there in our dress clothes, and ask the same of you.

As the date of the blockade approached, over a hundred groups lent their names in support, including peace groups, poverty groups, and environmental groups. So did a number of prominent individuals, including NASA's James Hansen; actors Mike Farrell, Daryl Hannah, Martin Sheen, and Susan Sarandon; musicians Will.I.Am, Goapele, John Densmore, and Kathy Mattea; writers Naomi Klein, David Korten, Noam Chomsky, and Paul Hawken; and environmentalists Paul Ehrlich and Gus Speth.

On the day before the action, a sudden blizzard hit the Chesapeake Bay region, dumping half a foot of wet snow on Washington and snarling traffic on highways and city streets. Organizers rushed to complete the last of hundreds of nonviolence trainings. Despite concerns that turnout would be decimated, a large and spirited crowd at an emergency planning meeting unanimously decided to press on with the action.

The next morning, an estimated four thousand protesters gathered at Liberty Park, then marched toward the Capitol Power Plant surrounded by red, green, blue, and yellow flags and banners and led by a contingent of Native American and Appalachian leaders. Reaching the plant, the demonstrators divided into four groups, each blocking one of the entrances. At each entrance, helmeted police guarded the gates, in effect enforcing the objective of the march to shut down the plant. Listening to speeches, chanting, and singing songs, protesters shivered in the 23-degree cold.

It was clear that the police were in "stand back" mode—accepting the blockade and intent on avoiding arrests. In effect, the demonstrators had won, though many were disappointed that the opportunity had been lost to dramatize the issue of coal worldwide through hundreds of arrests at the heart of the nation's capital city.

In fact, congressional leaders had already preempted the protest. During the week leading up to the blockade, House Speaker Nancy Pelosi and Senate Majority Leader Harry Reid had directed the Capitol architect, Stephen Ayers, to switch the Capitol Power Plant to natural gas. Two months later Ayers reported that the switch had been accomplished, though coal would be reserved for times of unusually cold weather or equipment outages.

Of course, the demonstration had never been just about the Capitol Power Plant, which was actually a fairly small facility compared to the typical coal-fired power plant. Among the other goals that the organizers had hoped to accomplish were to move coal policy into the center of the national conversation on climate, to push for stronger legislative action on climate, and to legitimize direct action as a movement tool.

Based on events over the subsequent months, the final goal showed the clearest results. Following the Capitol Climate Action, the number of direct action protests against coal immediately increased both in frequency and size. At the Cliffside Plant in North Carolina, hundreds protested and forty-eight were arrested. Another fourteen were arrested blockading TVA headquarters in Knoxville. In West Virginia, five activists were arrested unfurling a 40-foot-tall banner that read "EPA stop MTR" at Massey Energy's Edwight mine. In Nottingham, England, in a preemptive strike aimed at preventing a large direct action protest at the Ratcliff-on-Soar coal plant, police arrested 114 people at a community center and school. Around the world, more than two hundred people began fasting for up to forty days to draw attention to the urgency of global warming.

To provide a greater sense of coherence and planning to this ongoing swirl of protest, three dozen organizations had met in November to initiate the Power Past Coal campaign. The goal of the campaign, which kicked off on January 21, 2009, was to sponsor an action against coal on each of the first one hundred days of the Obama administration. By the halfway point of the campaign, the hundred-action goal had already been surpassed.

A glance through the list of actions that took place around the country reads like a catalog of the movement itself: "Dirty Movie Nights" in Oak Ridge, Tennessee; "Cliffside Boycott Party" in Asheville, North Carolina; "Valentine's Day Action for Black Mesa" in Flagstaff, Arizona; "NY Coal Trade Association Protest" in New York City.

In Boston, Massachusetts, a mannequin was found chained to the doors of the Kenmore Square Bank of America. Purportedly representing the group Mannequins for Climate Justice, the mannequin carried a notice reading "Even a dummy like me can see that Bank of America's massive loans to coal companies and support for the epidemic of foreclosures and evictions have to stop now."

Midway through the Power Past Coal campaign, at a crowded meeting in Washington, D.C., across the street from the PowerShift conference, organizers from around the country met in a brainstorming session to plan how to wrap up the campaign. As usual, the challenge faced by Dana Kuhnline of the Alliance for Appalachia, who along with intern Sierra Murdoch had spearheaded the campaign, was to raise the national visibility of a movement that was largely rural and widely dispersed. An answer was suggested by Marie Gladue Dine of the Black Mesa Water Coalition, who noted the tradition in some Native American religions of making offerings to the six directions: North, South, East, West, Sky, and Earth. The idea bore fruit on April 26, when six activists, each representing a strand of the anti-coal movement, returned to Washington to publicize the results of the Power Past Coal campaign, to lobby Congress, and to speak on behalf of thousands of individuals who had taken part in over three hundred actions in all fifty states. The spokespeople included L.J. Turner, a Wyoming rancher and member of the Western Organization of Resource Councils; Marie Gladue Dine of the Black Mesa Water Coalition in Arizona; Mike Cherin, an organizer with the Canary Coalition clean air advocacy group in North Carolina; Samuel Villaseñor, an organizer with the Little Village Environmental Justice Organization in Chicago; Towana Yepa, a member of the Little River Band of Ottawa Indians in Michigan; and Lorelei Scarbro, an organizer with Coal River Mountain Watch in West Virginia.

All six activists came from areas that had already been severely impacted by mines and power plants. All shared the common goal of putting a halt to destructive mining, initiating an orderly phase-out of the six hundred existing coal plants, and creating a new energy infrastructure based on efficiency and clean energy generation.

As I considered the nature of the challenge ahead, I noticed that British billionaire Sir Richard Branson, the founder of Virgin Airlines, had established a $25 million prize known as the Virgin Earth Challenge to the first inventor who figured out a way to remove 10 billion metric tons of carbon dioxide from the atmosphere. To me, Branson was missing the point entirely. The problem was not a lack of affordable technology. Wind and solar power were already available for commercial deployment, and the "intermittency problem" was already being addressed through new storage technologies and better integration of the grid.

Meanwhile, Europe, Japan, California, and other locations had already shown that energy efficiency standards and investments could reduce power consumption by half or more. In short, the problem wasn't technical; rather, it was political. As long as the coal industry remained politically dominant, there would be little point in inventing yet more alternative technologies if King Coal could simply find ways to block their implementation. More than new lab work, the real challenge of climate change lay in broadening the reach of grassroots organizing and political mobilization. Ultimately, humanity's fate would be decided not in the laboratories but in the streets, on campuses, on the steps of legislatures and courthouses, at the gates of factories and mines, in the doorways of banks and stock exchanges—anywhere people gathered and acted in concert to make change.

Two years earlier, the 151 proposed coal plants listed on Erik Shuster's spreadsheet had looked like nothing short of a planetary doomsday list. Any one of those coal plants, if built, would have emitted millions of tons of carbon dioxide each year. But in July 2009, the Sierra Club reported that 100 coal plants had been cancelled, and shortly after that the club added yet another cancellation to the list. Collectively, those 101 plants amounted to over 60,000 megawatts of generating capacity that could now be replaced with climate-friendly technologies. Assuming an average lifespan of fifty years, those 101 plants would have emitted 20 billion tons of carbon dioxide, twice the 10-billion-metric-ton goal of the Virgin Earth Challenge.

No doubt, when Branson devised the prize he was thinking about how to motivate the proverbial garage inventor or moonlighting chemist to come up with a new planet-rescuing technology in the narrow sense of the term—perhaps some sort of chemical reagent, gene-tweaked algae, or superabsorbent biochar that could suck carbon dioxide molecules out of the atmosphere. But if civilization is going to survive, it is time for visionaries like Branson to do some out-of-the-box thinking about technology itself, starting with the meaning of the term.

Wikipedia's definition of technology is as good as any:

> A strict definition is elusive; "technology" can refer to material objects of use to humanity, such as machines, hardware or utensils, but can also encompass broader themes, including systems, methods of organization, and techniques.

The "technologies" of grassroots politics used by the anti-coal movement—community organizing, non-violent direct action, corporate campaigning, Web 2.0 networking, regulatory intervention and litigation, etc.—are neither complex nor mysterious, but with them the movement has accomplished the astonishing feat of putting the brakes on a runaway train that promised to kill any hope of halting catastrophic climate change. If the movement had not challenged the wave of new plants, the vast majority would have been built, and the result would have been to lock the U.S. energy system into ever-rising emissions of greenhouse gases and undermine climate-safe investments.

The dynamics of plant cancellations are complex, typically amounting to a combination of factors that may include rising construction costs, legal challenges, public and political opposition, and regulatory delays. Grassroots action employs a wide variety of techniques—from sit-ins to press releases to legal briefs—to bring all the stars into alignment. There's a bit of alchemy involved, a bit of "fake it till you make it," and lots of sheer scrambling. Each situation is unique.

Obviously, Richard Branson is not about to write a check to the No New Coal Plants movement for $25 million. For starters, there is no organization called "No New Coal Plants Movement." The CoalSwarm Web site shows at least 250 separate organizations working to oppose coal plants and mines. But when one considers what that movement has accomplished on a shoestring, it is interesting to imagine what could be accomplished if Branson were to distribute the Virgin Earth Challenge prize among those groups.

Despite its accomplishments, the anti-coal movement continues to operate largely out of the public spotlight. But at least some observers seem to recognize its significance. Writing in the *Manchester Guardian*, British journalist Juliette Jowit reported:

> In a few years, the backlash against coal power in America has become the country's biggest-ever environmental campaign, transforming the nation's awareness of climate change and inspiring political leaders to take firmer action after years of doubt and delay. Plants have been defeated in at least 30 of the 50 states, uniting those with already

strong environmental records, such as California, with more conservative areas, such as the southern and central states.

Even Jowitt's description understates the movement's accomplishments. To me, the No New Coal Plants movement represents evidence that civilization as a whole—the planetary brain—might possess a quality that psychologists sometimes refer to as "executive function," the ability to prioritize one's actions and energies, focusing on the most important. At this point in history, climate change is generally recognized as the most important challenge facing humanity, an existential crisis for civilization itself. The scientists who have studied the problem for decades have concluded that ending emissions from coal is the key to heading off dangerous climate change. The fact that enough people have grasped the danger, focused on the solution, and joined effectively to accomplish political change—all this shows a civilization capable of thinking on its feet. The odds remain daunting, but this is reason for hope.

■

Protests Against Coal

■

2003

AUGUST **BLOCKADE AT ZEB MOUNTAIN.** On August 18, 2003, the Rocky Top Trio affinity group of Katúah Earth First! locked into concrete-filled steel barrels, blocking the entrance to the Zeb Mountain mine in Tennessee. The three protesters, john johnson, Dan Anderson, and Matthew Hamilton, were arrested and released that day. Near the mine on the same day, the Banner Busters affinity group climbed a nearby 150-foot billboard off Interstate 75 and hung a banner reading "Stop Mountaintop Removal."

2004

NOVEMBER **CHESAPEAKE CLIMATE ACTION NETWORK BLOCKADE OF DICKERSON POWER PLANT.** On November 10, 2004, a group of Chesapeake Climate Action Network activists, students, farmers, and religious officials held a protest against the coal-fired Dickerson Power Plant in Montgomery County, Maryland. During the protest, six people were arrested for blocking the entrance road to the plant. Protesters called on the plant's owner, the Mirant Corporation, to stop opposing state and federal legislation against power plant pollution.

2005

MARCH **SAVE HAPPY VALLEY COALITION OCCUPATION OF SOLID ENERGY HEADQUARTERS.** On March 6, 2005, four Save Happy Valley Coalition activists locked down at the corporate headquarters of Solid Energy in Christchurch, New Zealand, in protest of Solid Energy's plans to build a coal mine in Happy Valley. Supporters hung banners and pitched tents on Solid Energy's property. The occupation came one day after Solid Energy sued three activists for defamation.

JUNE **MOUNTAIN JUSTICE SUMMER PROTEST AT NATIONAL COAL CORPORATION.** On June 7, 2005, approximately forty-five Mountain Justice Summer activists, some in animal costumes, surprised the first-ever shareholders meeting of Knoxville-based National Coal Corporation with a marching band, chants, drumming, and noise makers. Demonstrators demanded that National Coal stop mountaintop removal mining and distributed informational fliers to shareholders. The sheriff and National Coal Corporation security personnel responded by assaulting protesters with pain compliance and choke holds, and they arrested three on felony charges.

WEST VIRGINIA CITIZENS OCCUPY MASSEY HEADQUARTERS. On June 30, 2005, concerned parents, grandparents, and other citizens of Coal River Valley, with support from Mountain Justice Summer participants, delivered a list of demands to Massey Energy's headquarters in Richmond, Virginia. Two were arrested for trespassing when they refused to leave the premises until Massey responded to their demands. Citizens demanded that Massey shut down its preparation plant, coal silo, 1,849-acre mountaintop removal coal mine, and 2.8-billion-gallon coal sludge dam located uphill from Marsh Fork Elementary School in Sundial, West Virginia.

JULY **FIRST NATIONS MOUNT KLAPPAN MINE BLOCKADE.** On July 16, 2005, representatives of three British Columbia First Nations tribes—the Telegraph Elders, the Tl'abânot'în Clan, and the Iskut First Nation—blockaded a road leading to the Mount Klappan coalfields in northwestern British Columbia. Tl'abânot'în tribe members had notified the mine's owners, Fortune Minerals, that their mine infringed upon Tl'abânot'în Aboriginal Title and Rights, as the company had failed to consult adequately with the tribe; Fortune Minerals had ignored the tribe's appeals. The blockade was maintained for seven weeks.

AUGUST **SAVE HAPPY VALLEY COALITION COAL TRAIN BLOCKADE.** On August 13, 2005, twenty-five Save Happy Valley Coalition activists and allies blockaded train tracks leading from Solid Energy's coal mines to the port of Lyttelton, New Zealand, in protest of Solid Energy's plans to build a coal mine in Happy Valley. Two people locked themselves to the tracks, while a third suspended himself from a tree a hundred feet in the air, attached to a support rope that was tied to the tracks. Four Solid Energy trains stood on the tracks for five hours while police cleared the blockade; the company claimed in court that the blockade cost it $150,000. The three blockaders were arrested.

EARTH FIRST! AND MOUNTAIN JUSTICE SUMMER BLOCKADE OF CAMPBELL COUNTY MOUNTAINTOP REMOVAL SITE. On August 15, 2005, Earth First! and Mountain Justice Summer activists blockaded a road leading to National Coal's mountaintop removal coal mine in Campbell County, Tennessee. Activists stopped a car on the road, removed its tires, locked themselves to the vehicle, and erected a tripod with a person perched on top of it. National Coal workers arrived and threatened protesters; one tried to ram the tripod with his car. Eleven people were arrested; the police treated the arrested activists roughly, endangering their safety.

2006

JUNE

RISING TIDE BOAT BLOCKADE OF NEWCASTLE, AUSTRALIA, PORT. On June 5, 2006, seventy people from Rising Tide used small boats to blockade the port of Newcastle, which exports 80 million tons of coal each year. The protest aimed to call attention to a planned expansion that would allow the port to export twice that amount.

JULY

EARTH FIRST!/RISING TIDE BLOCKADE OF CLINCH RIVER POWER PLANT. On July 10, 2006, seventy-five Earth First! and Rising Tide North America activists blockaded an access bridge leading to American Electric Power's coal-fired Clinch River Power Plant near Carbo, Virginia. Several people stretched a rope across the bridge and suspended themselves off the bridge's edge; others waved a coal truck onto the bridge, blockaded it, deflated its tires, and locked themselves to the truck. Protesters demanded that Clinch River and other outdated coal plants be shut down and that mountaintop removal coal mining be ended. After several hours during which coal trucks were unable to get into the plant, police agreed to make no arrests if the activists dismantled their blockades.

AUGUST

DRAX POWER STATION BLOCKADE ATTEMPT. On August 31, 2006, around six hundred people attempted to shut down the Drax Power Station in Selby, United Kingdom, in a widely publicized action that was organized by a variety of environmental groups and billed as "the battle of Drax." Several raiding parties of activists were arrested while trying to break through the perimeter fence. A larger crowd of people then pushed through police lines and were arrested as well. In a massive show of force, area police arrested thirty-eight people throughout the day. Many power plant staff didn't show up for the day, and others locked their doors.

DECEMBER **DOODÁ DESERT ROCK BLOCKADE.** On December 12, 2006, members of the Diné tribe blockaded a road leading to the planned site of the Desert Rock coal-fired power plant near Farmington, New Mexico, in protest of Sithe Global's failure to fully consult with members of the community. Ten activists with the group Doodá Desert Rock set up a campsite on the road. On December 22, under threat of arrest, the campsite was moved to a nearby location, and company vehicles were once again able to access the site. This second campsite was continually occupied for nearly a year. No arrests were made.

2007

FEBRUARY **RISING TIDE BLOCKADE OF NEW SOUTH WALES LABOR PARTY.** On February 27, 2007, fifteen Rising Tide Australia activists blockaded the headquarters of the New South Wales Labor Party in a protest of the provincial government's proposed plans to expand the Newcastle coal port. Activists blocked the door with several 44-gallon drums, and a woman chained herself to one of the blockades. They demanded that the provincial government announce whether or not the port would be expanded. Two people were arrested.

MARCH **SIT-IN AT WEST VIRGINIA GOVERNOR JOE MANCHIN'S OFFICE.** On March 16, 2007, dozens of West Virginia community members, together with activists from Mountain Justice Summer and Rising Tide North America, occupied the office of West Virginia governor Joe Manchin in protest of the State Mine Board's approval of construction permits for a second coal silo near Marsh Fork Elementary School in Sundial. Community activists demanded that the state move the school. Eleven people were arrested at this action, and many were treated roughly by police.

APRIL **BLOCKADE OF ASHEVILLE MERRILL LYNCH.** On April 13, 2007, two people calling themselves members of the "Climate Justice League" entered a Merrill Lynch building in Asheville, North Carolina, dumped a sack of coal in the lobby, and used a bicycle lock to blockade the door. They demanded that Merrill Lynch stop funding mountaintop removal coal mining companies such as Massey Energy. No arrests were reported.

JUNE **ASEN BLOCKADE OF NEW SOUTH WALES DEPARTMENT OF PLAN-NING.** On June 8, 2007, Australian Student Environment Network activists blockaded the office of the New South Wales Department of Planning. They criticized the department's June 7 decision to allow the Anvil Hill coal mine to fully drain the Hunter River in order to supply its mine with water. One person dressed as a polar bear chained herself to the doors of the building.

JULY **GREENPEACE BLOCKADE OF NEW SOUTH WALES DEPARTMENT OF PLANNING.** On July 3, 2007, Greenpeace Australia activists dumped four tons of coal in front of the door of the New South Wales Department of Planning, blocking the entrance to the building. They criticized the department's June 7 decision to allow the Anvil Hill coal mine to fully drain the Hunter River, in order to supply its mine with water. The sign outside the office was changed to read "Department of Coal Approvals." No arrests were reported.

AUGUST **SOUTHEAST CONVERGENCE FOR CLIMATE ACTION OCCUPATION OF ASHEVILLE BANK OF AMERICA.** On August 13, 2007, 150 activists from Southeast Convergence for Climate Action occupied a Bank of America branch in Asheville, North Carolina. They condemned Bank of America's ongoing funding of mountaintop removal mining in Appalachia. Two people locked themselves to the main lobby, while others blockaded the entrance to the branch and delivered coal to the bank's managers. Five people were arrested.

SEPTEMBER **OCCUPATION OF LOY YANG POWER PLANT.** On September 3, 2007, activists from Real Action on Climate Change occupied the coal fired Loy Yang Power Station in Traralgon, Australia. Two people chained themselves to the coal conveyor belt and others hung several large banners from the plant. The action, which took place several days before an Asia-Pacific Economic Cooperation summit in Sydney, was intended to draw attention to Prime Minister John Howard's failure to limit Australian carbon emissions. Four people were arrested.

ASEN OCCUPATION OF NEWCASTLE COAL PORT. On September 4, 2007, twenty activists from the Australian Student Environment Network occupied the coal port in Newcastle, Australia. Five people chained themselves to machinery at the Carrington Coal Terminal. The action took place several days before an Asia-Pacific Economic

Cooperation summit in Sydney; it was intended to draw attention to Prime Minister John Howard's failure to limit Australian carbon emissions. Eleven people were arrested.

OCTOBER **GREENPEACE OCCUPATION AT BOXBURG PLANT CONSTRUCTION SITE.** Beginning October 1, 2007, thirty-four activists occupied the construction site of a new coal-fired power plant in Boxburg, in eastern Germany. The activists, ten of whom remained camped atop cranes on the site for sixty hours, demanded that Vattenfall, the utility sponsoring the plant, stop building coal plants and instead invest in renewable energy. A giant banner hung from a crane read "Vattenfall: Stop building! Climate protection instead of brown coal!" Six smaller banners reading "Stop CO_2" hung from other cranes. Volunteers painted "Stop CO_2" onto a smokestack under construction.

GREENPEACE OCCUPATION OF KINGSNORTH POWER PLANT. On October 8, 2007, fifty Greenpeace UK activists occupied the Kingsnorth Power Station near Kent, United Kingdom. One team of people shut down the conveyor belts carrying coal into the plant and then chained themselves to the machinery. Another team scaled the plant's chimney, upon which they painted the phrase "Gordon Bit It." Greenpeace held the action to protest plans by the plant's owner, E.ON, to build two new coal-fired plants at the site, which would be the first coal-fired power plants built in the United Kingdom in twenty years. Police arrested eighteen people during the action.

RAINFOREST ACTION NETWORK BANNER HANG AT BANK OF AMERICA CORPORATE HEADQUARTERS. On October 23, four activists with Rainforest Action Network scaled a fifteen-story crane across the street from Bank of America's corporate headquarters in downtown Charlotte, North Carolina. Reading "Bank of America: Funding Coal, Killing Communities," the banner hang protested the bank's funding of mountaintop removal and new coal plant development. The banner hang disrupted traffic for several blocks until police and firefighters brought down the activists. All four were arrested.

NOVEMBER **RISING TIDE BOAT BLOCKADE OF NEWCASTLE PORT.** On November 3, 2007, a hundred people from Rising Tide again blockaded the port of Newcastle, Australia, which exports 80 million tons of coal each year. The protest aimed to call attention to a planned expansion that would allow the port to double the tonnage exported. Participants attempted to block ships from entering the port for four hours, but police boats managed to escort three ships into the port. At one point, a police Jet Ski rammed a woman's kayak, resulting in her hospitalization.

RAINFOREST ACTION NETWORK ACTIVISTS AND ALLIES BLOCKADE A CITIBANK BRANCH IN WASHINGTON, D.C. On November 5, 2007, activists from Rainforest Action Network, Coal River Mountain Watch, and the Student Environmental Action Coalition joined hundreds of student activists in blockading a Citibank branch in Washington, D.C., to protest Citibank's ongoing funding of new coal power plant development. RAN activists performed a "die-in" and delivered a wheelbarrow full of coal to the bank's managers. Police shut the branch down for the day, and no arrests were made.

RAINFOREST ACTION NETWORK DAY OF ACTION AGAINST COAL FINANCE. On November 15, 2007, Rainforest Action Network activists—acting with allies from Coal River Mountain Watch, Appalachian Voices, Rising Tide North America, Mountain Justice Summer, Student Environmental Action Coalition, and Energy Justice Network—staged dozens of actions against Citibank and Bank of America branches in cities across the country in protest of the two companies' refusal to stop funding new coal power plant development and coal mountaintop removal mining. In San Francisco, RAN activists attached caution tape—reading "Global Warming Crime Scene"—to dozens of Bank of America and Citibank ATMs and held "cough-ins" in several branches. Similar ATM closure actions were held in New York City; Davis, California; Los Angeles, California; Portland, Oregon; and St. Petersburg, Florida. Protests against the two companies were held in numerous other cities.

STUDENT BLOCKADE OF DUKE ENERGY HEADQUARTERS. On November 15, 2007, two Warren Wilson College students—dressed as polar bears—chained themselves to the door of Duke Energy's headquarters in Charlotte, North Carolina, in protest of Duke's plans to build the Cliffside coal-fired power plant in western North Carolina. Several dozen people held a rally in support of their blockade, dressing as Santa Claus and elves and presenting a stocking full of coal to the company. The two students were arrested on charges of trespassing and disorderly conduct.

GREENPEACE OCCUPATION OF MUNMORAH POWER STATION. On November 15, 2007, fifteen Greenpeace Australia activists occupied the Munmorah coal-fired power plant near Wyong, Australia. Two teams of five people—including engineers—switched off the conveyor belt that brings coal into the plant and then chained themselves to the machinery. Another team painted "Coal Kills" on the roof of the plant and hung a large banner inside. The action took place several days before Australian parliamentary elections; it was held in protest of the climate change policies of both major Australian political parties. Police arrested all fifteen people.

RISING TIDE KOORAGANG COAL TERMINAL RAIL BLOCKADE. On November 19, 2007, several Rising Tide Australia activists blocked a train carrying coal to the Kooragang Island coal terminal, from which 80 million tons of coal are exported each year. One person chained himself to the train; he was later arrested. Protesters demanded that the Australian government begin to reduce the country's reliance on coal.

DECEMBER **BLOCKADE OF FFOS-Y-FRAN COAL MINE CONSTRUCTION SITE.** On December 5, 2007, about thirty local residents and activists from a variety of environmental groups—many dressed as polar bears—occupied the Ffos-y-fran coal mine construction site in South Wales, being built about forty yards from several homes. Activists dressed as polar bears chained themselves to bulldozers, while other people hung a banner from one bulldozer criticizing Prime Minister Gordon Brown's ongoing support for coal power. The action was timed to coincide with the Bali climate change negotiations.

2008

MARCH **MOUNTAIN JUSTICE SPRING BREAK ACTION AT AMP-OHIO HEADQUARTERS, COLUMBUS, OHIO.** On March 28, 2008, activists participating in Mountain Justice Spring Break occupied the lobby of American Municipal Power–Ohio's headquarters in Columbus and demanded a meeting with AMP's CEO Marc Gerken. Several people stated their intention to conduct a sit-in in the office if their demands weren't met; about forty people protested outside. After thirty minutes, Gerken met with protesters and agreed to schedule a meeting of the Board of Trustees at which community members could present their concerns with AMP-Ohio's proposed coal-fired power plant in Meigs County, Ohio. No arrests were made.

APRIL **RISING TIDE AND EARTH FIRST! OCCUPATION OF CLIFFSIDE CON-STRUCTION SITE.** On April 1, 2008, as part of the Fossil Fools International Day of Action, a group of North Carolina activists with Rising Tide and Earth First! locked themselves to bulldozers to prevent the construction of the Cliffside coal-fired power plant proposed by Dominion in western North Carolina. Others roped off the site with "Global Warming Crime Scene" tape and held banners protesting the construction of the plant. Police used pain compliance holds and tasers to force the activists to unlock themselves from the construction equipment. Eight people were arrested.

RAINFOREST ACTION NETWORK BLOCKADE OF A CITIBANK OFFICE IN NEW YORK CITY. On April 1, 2008, as part of the Fossil Fools International Day of Action, twenty-five Billionaires for Coal blockaded Citibank's Upper West Side headquarters in New York City. Two people chained themselves to the door, while others—dressed in tuxedos and top hats—drew attention to Citibank's funding of new coal power plant development and mountaintop removal mining. Police cut through the chains locking the two billionaires to Citibank's door and arrested them.

RISING TIDE AND RAINFOREST ACTION NETWORK BLOCKADE OF BOSTON BANK OF AMERICA BRANCH. On April 1, 2008, as part of the Fossil Fools International Day of Action, four activists used lockboxes to block the entrance to a Bank of America branch in Boston, in protest of BofA's investments in mountaintop removal mining and new coal power plant development. Others held banners and signs in support of the action, which was organized by Rising Tide North America and Rainforest Action Network. Police used saws to cut through the lockboxes and arrested the four blockaders.

OCCUPATION OF FFOS-Y-FRAN COAL MINE CONSTRUCTION SITE. On April 1, 2008, as part of the Fossil Fools International Day of Action, dozens of local residents and activists from a variety of environmental groups occupied the Ffos-y-fran coal mine construction site in South Wales. Protesters arrived at 6 a.m., scaled a coal washery and dropped a 100-foot banner, took over construction machinery, and locked themselves to the front gate, shutting down major work at the site for the day. Police made two arrests, and the other activists left without incident.

EASTSIDE CLIMATE ACTION BLOCKADE OF E.ON HEADQUARTERS, NOTTINGHAM, UNITED KINGDOM. On April 1, 2008, as part of the Fossil Fools International Day of Action, thirty activists with Eastside Climate Action blockaded the front entrance of E.ON UK's headquarters in Nottingham. Two people used U-locks to lock themselves to the front door, while others blockaded the back entrance; other protesters poured green paint on themselves to simulate E.ON's "greenwashing." The action was in protest of E.ON's plans to build the Kingsnorth coal-fired power plant, the first new coal plant in the United Kingdom in fifty years. Police made two arrests, and the building was shut down for the day.

RISING TIDE OCCUPATION OF ABERTHAW POWER STATION. On April 3, 2008, as part of the Fossil Fools International Day of Action, members of Bristol Rising Tide occupied the Aberthaw coal-fired power plant, operated by RWE Power in South Wales. Activists entered the facility, chained themselves to conveyor belts, and occupied several buildings; others locked themselves to the facility's front gates. The action was in solidarity with the protests at the Ffos-y-fran mine construction site in South Wales; coal from Ffos-y-fran will be used to fuel Aberthaw for seventeen years. Police arrested eleven people.

BLUE RIDGE EARTH FIRST! BLOCKADES DOMINION POWER'S HEADQUARTERS. On April 15, 2008, fifteen activists with Blue Ridge Earth First! blockaded the entrance of Dominion Power's headquarters to protest Dominion's planned coal-fired power plant in Wise County. Three activists locked themselves to trash cans filled with concrete and blocked both lanes of the only road in and out of the office complex. The blockade, established just before 8 a.m., held for almost two hours and backed up traffic almost a mile. The locked-down activists were eventually dragged to the side of the road by police and given citations for impeding the flow of traffic.

RISING TIDE BLOCKADE OF COAL TERMINAL CONSTRUCTION SITE IN NEW SOUTH WALES. On April 19, 2008, fifty Rising Tide Australia activists stormed the gates of a coal terminal construction site in Newcastle, New South Wales. Once inside, about twenty of the protesters locked arms and refused to leave; eighteen were arrested. They were protesting the planned expansion of the facility.

JUNE   **ACTIVISTS HALT COAL TRAIN ON ITS WAY TO UNITED KINGDOM'S LARGEST POWER PLANT.** On the morning of June 13, 2008, forty Camp for Climate Action activists, a small number disguised as railway workers, flagged down and stopped a coal train on its way to Drax Power Station, the United Kingdom's largest power plant. Some protesters climbed onto the train and unloaded almost 20 tons of coal onto the tracks, while others chained themselves to the train. A banner was unfurled reading "Leave It in the Ground!" Riot police stormed the train and removed protesters around midnight, arresting twenty-nine.

PROTESTERS UPSTAGE BRISBANE COAL CONFERENCE. Protesters rallied outside while two campaigners infiltrated a major coal conference in Brisbane, Australia. Once inside, the two activists took the floor and addressed the Queensland Coal08 conference, which was held to discuss the future of the coal-mining industry in the largest coal-exporting state in the largest coal-exporting country in the world. No arrests were made.

ACTIVISTS DEMONSTRATE OUTSIDE BANK OF AMERICA HEADQUARTERS. On June 26, 2008, activists from Rainforest Action Network demonstrated outside Bank of America's headquarters in Charlotte, North Carolina, carrying a banner that read "Divest from Coal!" The group distributed fliers to employees about the bank's investments in the coal industry and local residents. Police were on hand, but no one was arrested.

ACTIVISTS BLOCKADE DOMINION HEADQUARTERS. On June 30, 2008, twenty activists with Blue Ridge Earth First! and Mountain Justice Summer blockaded the entrance to Dominion's corporate headquarters to protest the company's plan for the new coal-fired Wise County Plant in southwest Virginia. Four protesters formed a human chain with their hands encased in containers of hardened cement and a fifth dangled by a climber's harness from the Lee Bridge footbridge. After several hours police made their way through miles of backed-up traffic to cut the activists out of the lockboxes and barrels. The climber came down on his own. Police also detained eight others standing on the sidewalks supporting the lockdown team. Thirteen people were arrested.

JULY **GREENPEACE ACTIVISTS SHUT DOWN A PORTION OF AUSTRALIA'S MOST POLLUTING POWER STATION.** At dawn on July 3, 2008, twenty-seven Greenpeace activists entered the 2,640-megawatt Eraring Power Station site north of Sydney to call for an energy revolution and the end of coal. Twelve protesters shut down and chained themselves to conveyors while others climbed onto the roof to paint "Revolution" and unfurled a banner reading "Energy Revolution—Renewables Not Coal." The action preceded the delivery by Australian climate change advisor Professor Ross Garnaut of his Draft Climate Change Review on July 4. Police arrested twenty-seven people. Eraring Power Station, near Sydney, releases nearly 20 million metric tons of greenhouse pollution into the atmosphere every year.

EARTH FIRST! ACTIVISTS LOCK DOWN AT AMERICAN MUNICIPAL POWER HEADQUARTERS, COLUMBUS, OHIO. On July 7, 2008, approximately seventy-five Earth First! activists gathered outside American Municipal Power (AMP) headquarters in Columbus to protest the company's plan to build the new 960-megawatt coal-fired American Municipal Power Generating Station in Meigs County, Ohio. Two protesters climbed flagpoles in front of the building and hoisted banners that read "No New Coal!" and "We won't stop until you do." Around twenty activists entered the building and occupied the lobby as five protesters connected themselves to each other using lockboxes. Police used pepper spray on protesters and arrested eight when they refused to leave.

MOUNTAIN JUSTICE ACTIVISTS PROTEST APPROVAL OF COAL GAS-IFICATION PLANT, BOSTON, MASSACHUSETTS. On July 10, 2008, nearly fifty Mountain Justice Summer activists gathered in opposition to a coal project in Massachusetts, donning haz-mat suits and delivering a pile of coal while displaying "Global Warming Crime Scene" caution tape on the front steps of the Office of Energy and Environmental Affairs in Boston. The action was in response to the office dismissing an appeal of the state's approval for a coal gasification project in Somerset, Massachusetts.

GREENPEACE ACTIVISTS OCCUPY COAL-FIRED POWER PLANT SMOKE-STACK FOR THIRTY-THREE HOURS. On July 11, 2008, four Greenpeace activists climbed the 462-foot smokestack of the Swanbank power station near Brisbane, Queensland, Australia. While the smokestack climbers hung a "Renewables Not Coal" banner, two other activists climbed onto the roof of the plant and unfurled a banner reading "Energy [R]evolution." The four remained on the smokestack overnight in near-freezing temperatures. On July 12, one of the protesters painted "Go Solar" in huge lettering down the side of the smokestack. After thirty-three hours of occupation, all four climbers descended voluntarily.

AUSTRALIA CLIMATE CAMP STOPS COAL TRAINS AT WORLD'S LARGEST COAL EXPORT PORT. On July 13, 2008, approximately a thousand activists stopped three trains bound for export at the Carrington coal terminal in Newcastle, Australia, for almost six hours. Dozens of protesters were able to board and chain themselves to the trains while others lay across the tracks. Hundreds were held back by mounted police. Police arrested fifty-seven. The actions were organized as part of the Australian Camp for Climate Action.

BLOCKADES AT KOORAGANG AND CARRINGTON COAL TERMINALS. On July 14, 2008, five activists stopped coal loading at the Kooragang coal terminal for more than two hours by chaining themselves to a conveyor belt. Later that afternoon four protesters padlocked themselves to the tracks at the Carrington coal terminal, stopping all train traffic until police were able cut the group free. All nine were arrested. The direct actions, organized as part of the Australian Camp for Climate Action, were an attempt to bring worldwide attention to coal's role in climate change and the expansion of Australian coal exports.

UK ACTIVISTS TARGET COAL-FIRED PLANT'S PR AGENCY. On July 16, 2008, activists with Oxford Climate Action blockaded the headquarters of public relations giant Edelman Public Relations. Several protesters gained access to the firm's offices while others climbed onto the roof to unfurl a banner reading "Edelman: Spinning the Climate Out of Control." Edelman provides public relations services for E.ON, the world's largest investor-owned energy service provider. E.ON UK is proposing to upgrade its coal-fired Kingsnorth Power Station to use supercritical coal technology. Kingsnorth is currently considered to be a conventional coal plant, but under the European Union's Large Combustion Plant Directive, the plant would eventually have to be closed without the upgrade. According to activists, Edelman PR is engaging in a campaign to "greenwash" E.ON's continued investment in burning coal.

FOUR ARRESTED AT TENNESSEE STRIP MINE. On July 20, 2008, residents from coal-impacted communities throughout Appalachia gathered for a march at Zeb Mountain, the largest surface coal mine in Tennessee. The march, organized by United Mountain Defense, Mountain Justice Summer, and Three Rivers Earth First!, included political theater, life-sized puppets, and speeches. In an act of civil disobedience, four citizen activists walked across a line marked with police tape designating National Coal Corporation's property. The four were arrested without incident.

AUSTRALIAN CITIZENS BLOCKADE FARM TO STOP COAL EXPLORATION. On July 21, 2008, nearly two hundred residents and landowners in northern New South Wales blockaded a farmer's driveway to prevent a BHP Billiton drilling rig from entering the property to explore for coal deposits. Local residents are asking for an independent study into the effects of exploration and coal mining on underground water reserves. A court had previously issued an injunction against the landowner when he drove a grader across his driveway to prevent the exploratory team from entering his property.

GREENPEACE PAINTS ANTI-COAL MESSAGES ON TWENTY COAL SHIPS. Using inflatable rafts, nine Greenpeace activists painted anti-coal messages on twenty coal ships waiting to enter the world largest coal export port in Queensland, Australia. The action was intended to highlight the contradiction between the Australian prime minister's stated goals of reducing greenhouse pollution and doubling Australia's coal exports. All nine activists were arrested.

AUGUST **ACTIVISTS GLUE THEMSELVES TO COAL GIANT'S HEADQUARTERS.** On August 11, 2008, nine activists glued themselves to the revolving door and windows at BHP Billiton's headquarters in central London. The protesters also scattered coal across the floor of the lobby. According to one activist, the protest was to highlight that the "expansion of the coal industry is unacceptable in the face of impending climate chaos." The protest ended peacefully after ninety minutes and there were no arrests.

SOUTHEAST CONVERGENCE FOR CLIMATE ACTION LOCKS DOWN AT BANK OF AMERICA, RICHMOND, VIRGINIA. On August 11, 2008, fifty activists began marching at Monroe Park around noon and made stops at the offices of coal-mining giant Massey Energy, Virginia's Department of Environmental Quality, and Dominion Virginia Power and ended at Bank of America, a major funder of coal. Two activists were arrested after locking themselves to a Bank of America sign. The march and lockdown culminated a week of environmental and climate justice training, networking, and strategizing at the Southeast Convergence for Climate Action. The march included jesters, larger-than-life puppets, banners, and signs to raise awareness about the climate crisis.

SEPTEMBER **GREENPEACE RAINBOW WARRIOR LAUNCHES "QUIT COAL" PROTEST CAMPAIGN IN ISRAEL.** On September 9, 2008, two Greenpeace activists painted "Quit Coal" in English and Hebrew on the hull of a ship unloading coal at the Ashkelon power plant. The action was in opposition to the Israeli government's plan to build a new coal power plant in Ashkelon. Police with support from the Israeli navy arrested the captain, crew, and passengers of the *Rainbow Warrior.*

SEPTEMBER **TWENTY PROTESTERS LOCK DOWN AT DOMINION COAL PLANT CONSTRUCTION SITE IN WISE COUNTY, VIRGINIA.** In the early morning of September 15, 2008, around fifty protesters entered the construction site of Dominion Virginia's coal-fired Wise County Plant. Twenty protesters locked themselves to eight large steel drums, two of which have operational solar panels affixed to the top illuminating a banner reading "Renewable jobs to renew Appalachia." In addition to those locked to the construction site, over twenty-five protesters from across the country convened in front of the plant singing and holding a 10-by-30-foot banner, which said, "We demand a clean energy future." Police arrested eleven people. On the same day, in San Francisco, activists with Rainforest Action Network infiltrated Dominion CEO Thomas F. Farrell's presentation at Bank of America's Annual Investment Conference. Farrell's PowerPoint presentation was replaced with a slideshow of the Wise County Plant protest.

PRIME MINISTER'S OFFICE OCCUPIED. On September 15, 2008, constituents occupied Australian prime minister Kevin Rudd's Brisbane electorate office, staging a peaceful sit-in for several hours and demanding a discussion on the government's lack of response to proposals for phasing out of the coal industry. The action was the first in a week of national climate emergency protest events, which targeted the Queensland government and coal-mining corporations.

PROTESTERS SHUT DOWN A CITIBANK BRANCH IN CAMBRIDGE, MASSACHUSETTS. On September 27, 2008, students, members of community groups, and climate activists held a public rally outside Bank of America's Harvard Square branch, protesting both Bank of America's and Citibank's risky investment strategies, which have contributed to the current economic crisis and are jeopardizing the global climate. Demonstrators then marched to a nearby Citibank branch, where four activists wearing T-shirts reading "Not with Our Money" locked themselves to the entrance.

OCTOBER   **GREENPEACE "QUIT COAL" TOUR VISITS SPAIN, BOARDS COAL SHIP.** On October 6, 2008, four Greenpeace activists boarded a cargo ship importing coal from Colombia into Spain. Others painted "Quit Coal" in English and Spanish on the ship. The action was in protest of the Spanish government's heavy reliance on coal for the country's energy supply and its subsidies to the coal industry.

CITIZENS RALLY AT STATE CAPITOL AGAINST NEW COAL USE, LITTLE ROCK. On October 18, 2008, citizens from across the state of Arkansas rallied at the state capitol building in Little Rock to protest two new coal-fired power plants proposed for the state. Protesters asked for investment in wind energy and a ban on new coal plants.

PREMIER OF QUEENSLAND'S OFFICE OCCUPIED. On October 31, 2008, the community group Friends of Felton occupied Premier of Queensland Anna Bligh's office. The twenty-five participants demanded legislation to protect farmland from mining. The action was promoted as "Lunch with Anna," and outside the office a mock lunch of coal and polluted water was served to a Bligh impersonator. Friends of Felton formed after Ambre Energy announced plans to build a "clean coal" gasification plant and open pit mine.

ZOMBIE MARCH ON TOP COAL INVESTORS, BOSTON, MASSACHU-SETTS. On Halloween, zombies descended on Copley Square to visit local Bank of America and Citibank branches to protest their funding for new coal power plants. The action was organized by Rising Tide Boston. Similar events were held in North Carolina and California.

NOVEMBER **RISING TIDE ACTIVISTS SHUT DOWN BAYSWATER POWER STATION, NEW SOUTH WALES.** On November 1, 2008, a large group of people from Rising Tide Newcastle walked onto the site of Bayswater Power Station, the biggest source of greenhouse gas pollution in Australia. Four people locked onto both conveyors, shutting down coal input into the station for six hours. An additional twenty-five people walked onto the coal piles outside the power station, disrupting operations, and were arrested for trespass. The group called on the government to begin phasing out coal as quickly as possible, peaking carbon emissions by 2010 and taking the strongest possible position to the United Nations Council of Parties (COP) negotiations in Poznan and Copenhagen.

ACTIVISTS SHUT DOWN COLLIE POWER STATION, WESTERN AUS-TRALIA. On November 5, 2008, two activists chained themselves onto a conveyor belt at Collie Power Station, which produces 300 megawatts of Western Australia's electricity and consumes around a million metric tons of coal per year. Lee Bell, a spokesperson for the group, said that the protest was part of nationwide action against the government's inaction on climate change and the failure to phase out coal-fired power.

ACTIVISTS SHUT DOWN HAZELWOOD POWER STATION. On November 6, 2008, a group of activists walked onto the site of the Hazelwood power station, one of the most inefficient power stations in the industrialized world, to protest Australian inaction on climate change. Two people chained themselves to the conveyor belts that carry coal to the power station. The station was due to be decommissioned in 2009 but instead is undergoing rapid expansion.

ACTIVISTS SHUT DOWN TARONG POWER STATION, QUEENSLAND, AUSTRALIA. On November 7, 2008, two activists locked onto a conveyor belt and forced the evacuation of Queensland's 1,400-megawatt Tarong Power Station. The action was the fourth in seven days targeting the coal industry in Australia and calling for the phaseout of coal-fired power. The action also served to highlight the risk to Queensland's world heritage icon, the Great Barrier Reef, posed by climate change. Three people were arrested.

NATIONAL DAY OF ACTION AGAINST COAL FINANCE (NOVEMBER 14–15, 2008). Thousands of activists around the United States mobilized to protest coal mining, coal-fueled power plants, and coal financiers. Groups involved in the action included Rainforest Action Network, Greenpeace, Rising Tide, Mountain Justice, Student Environmental Action Coalition, Coal River Mountain Watch, Ohio Valley Environmental Coalition, the Southern Energy Network, and Earth First! Activists placed anti-coal banners in strategic locations across the country, protested at Bank of America and Citibank branches, shut down ATMs with crime scene tape, and infiltrated Bank of America's Energy Conference.

GREENPEACE ACTIVISTS PROTEST OUTSIDE MINE, POZNAN, POLAND. On November 25, 2008, about two dozen Greenpeace activists protested at a new opencast mine and waved "Quit Coal!" banners before being forcefully removed from the area by miners. The incident drew attention to the United Nations Council of Parties on climate change, held in the city of Poznan.

ACTIVIST SHUTS DOWN KINGSNORTH POWER STATION IN THE UNITED KINGDOM. On November 28, 2008, in full view of security cameras, a single activist climbed two 10-foot razor-wired and electrified security fences at E.ON's coal-fired power plant and crashed a huge 500-megawatt turbine, leaving behind a banner that read "No New Coal." The plant was down for four hours, cutting the United Kingdom's CO_2 emissions during the outage by an estimated 2 percent. Police were unable to find the perpetrator of the outage.

DECEMBER **SANTA PROTEST AT TENNESSEE VALLEY AUTHORITY HEADQUARTERS IN KNOXVILLE, TENNESSEE.** On December 5, 2008, with help from United Mountain Defense and Three Rivers Earth First! Santa Claus and his elves came armed with coal and switches for the largest purchaser of coal in North America: TVA. Santa read letters from sad children who could not go outside and play sometimes because of days when it is literally unhealthy to breathe in Knoxville, letters from children sad that their grandparents are dying slow deaths of extended asphyxiation while lugging around bottled oxygen, and letters from children complaining that mountains are being blown up to get at that coal. The children said they felt that the drinking water was important and that they liked playing in the forest. After being asked to leave the premises, the North Pole-based environmental group proceeded outside to sing anti-coal carols and hand out information sheets.

SANTA DETAILED AT TENNESSEE VALLEY AUTHORITY OFFICES IN CHATANOOGA, TENNESSEE. On December 11, while attempting again to deliver letters from sad children, Santa was detained by the TVA police for an hour and half and issued a warning citation for supposedly disrupting a board meeting which had officially ended. The arresting TVA officer became concerned when he discovered that Santa had switches concealed in his britches. Santa was released after being detained without milk and cookies. Santa told reporters: "I am depending on all the little activist elves to deliver more coal to federal agencies in hopes to influence the first 100 days of president-elect Obama's administration through the newly appointed agency heads. This new administration must make stopping strip mining and addressing the destructive impact of coal on Santa's children its first priority. Ho Ho Ho." At 4 p.m. on December 12, while Santa and his elves were dancing and singing, TVA sent out one of its head PR people, Gill Francis. Mr. Francis wanted to meet and negotiate with Santa but Santa was too busy and took a number. After finishing the dance, Santa had his head elf call Mr. Francis to come back out and negotiate. When Mr. Francis appeared, slightly out of breath, Santa said he was sorry and put coal and switches in Mr. Francis hands saying, "This is the least favorite part of my job Mr. Francis—but TVA has been veerrrrry naughty." As Mr. Francis stormed off, Santa and his elves resumed dancing.

2009

JANUARY   **SLUDGE SAFETY LOBBY DAY, CHARLESTON, WEST VIRGINIA.** On January 31, 2009, residents of southern West Virginia descended on the state capitol, bringing along jars of black water taken from their wells in Boone and Mingo counties. They spent the day lobbying legislators to stop slurry injections into sludge ponds until studies could show what toxic materials the slurries contain.

FEBRUARY **COAL RIVER MOUNTAIN ACTIVISTS ARRESTED, PETTUS, WEST VIRGINIA.** On February 3, 2009, five Coal River Mountain Watch activists were arrested and charged with trespassing after locking themselves to a bulldozer and a backhoe at a Massey Energy mountaintop removal site. The activists planted a banner for the Coal River Wind Project in protest of the impending 6,600-acre mountaintop removal mine. Later in the day, eight more activists were arrested during a demonstration against Massey Energy's preparations to blast the mountain. Environmentalists contend that the mountain would be better used for a wind energy project and that the blasting could destabilize the world's largest toxic coal slurry impoundment.

RISING TIDE BOSTON CRASHES ARCH COAL CEO LECTURE, CAMBRIDGE, MASSACHUSETTS. On February 5, 2009, seven activists from Rising Tide Boston disrupted a lecture given by Arch Coal CEO Steve Leer at Harvard University. Leer was speaking about the future of "clean coal" technology. The activists interjected information on the impacts of coal extraction, including their final question, "What gives you the right to gamble the future of civilization on a magic technology that doesn't exist?" While Leer ignored the question, two members of Rising Tide carried a banner on stage that read "The coal bubble is bursting—clean coal is a dirty lie." The lecture was funded by Bank of America, the single largest financial backer of mountaintop removal.

BILLIONAIRES FOR COAL VISIT DOMINION HEADQUARTERS IN RICHMOND, VIRGINIA. On February 7, 2009, about two dozen people identifying themselves as Billionaires for Coal gathered outside the headquarters of Dominion to lampoon the coal industry. The activists wore formal dress and sipped from wine glasses, while shouting pro-coal, anti-environment slogans including "Up with sea levels, up with profits." Bluegrass musicians also performed, calling themselves the We Love Money String Band. Although the group's signs and chants stayed on message with the billionaire facade, the activists distributed leaflets revealing that the demonstration was organized by Blue Ridge Earth First!

GRASSROOTS EFFORTS FORCE RADIO HOST ED SCHULTZ TO CONSIDER ANTI-COAL VIEWPOINTS. On February 6, 2009, radio host Ed Schultz interviewed Joe Lucas, senior vice president of communications for the American Coalition for Clean Coal Electricity. After three and a half hours of grassroots pressure through e-mails and phone calls, Schultz agreed to invite an anti-coal guest on his show. Schultz is admittedly pro-coal, but he acknowledged the pressure he was under to provide the other side of the story.

RESIDENTS PROTEST PROPOSED SANTEE COOPER PLANT, FLORENCE COUNTY, SOUTH CAROLINA. On February 12, 2009, more than a hundred residents of Florence County brought an inflatable smokestack to the courthouse to protest the permit that was granted to Santee Cooper to build the Pee Dee Generating Facility on the banks of the Great Pee Dee River. The plant would emit over 11 million tons of carbon dioxide per year, as well as sixty different toxic pollutants, including arsenic, dioxins, heavy metals, mercury, and selenium.

ACTIVISTS CLOSE ACCOUNTS WITH BANK OF AMERICA, SAN FRANCISCO, CALIFORNIA. On Valentine's Day, February 14, 2009, more than twenty-five activists from Rising Tide Bay Area in San Francisco served Bank of America a "foreclosure notice" for "failing to pay its social and environmental debts." Activists closed accounts with the bank, pulling out over $10,000. The action was part of a nationwide campaign against Bank of America organized by Rising Tide North America.

TWO ARRESTED FOR HALTING BLASTING AT MOUNTAINTOP REMOVAL SITE, RALEIGH COUNTY, WEST VIRGINIA. On February 16, 2009, two protesters were arrested for interfering with mountaintop removal blasting on Massey Energy's Edwight mine site near the Shumate sludge dam in Raleigh County. The Shumate sludge dam holds back 2.8 billion gallons of toxic sludge, the waste by-product of chemically cleaning coal, and sits directly above the Marsh Fork Elementary School.

HUNDREDS GATHER FOR COAL PROTEST IN FRANKFURT, KENTUCKY. On February 17, 2009, hundreds of activists from ILoveMountains and Kentuckians for the Commonwealth, as well as actress Ashley Judd, gathered outside the state capitol building to protest mountaintop removal mining and rally for proposed legislation that had been stuck for several years in the House Natural Resources and Environment Committee. The bill, sponsored by Congressman Don Pasley (D-Winchester), would prohibit mining operations from dumping refuse into adjacent streams, but coal interests in the legislature had managed to keep the bill from getting a vote on the floor.

MARCH IN CORPUS CHRISTI, TEXAS. On February 19, 2009, over two hundred citizens wearing respirators marched along the Corpus

Christi bay front to protest the proposed Las Brisas Energy Center. The marchers included local doctors, who warned that the plant would worsen asthma rates, heart attacks, cancer, neurological and behavioral problems, and failed births. Estimates suggest that the plant would produce over 21,000 tons of air pollution a year, more than the annual emissions of all the surrounding counties combined.

MARCH   **ACTIVISTS RALLY AGAINST COAL IN MASSACHUSETTS.** On March 1, 2009, citizens across Massachusetts rallied outside the state's three major coal plants to show support for the Capitol Climate Action protest in Washington, D.C. The largest demonstration was in Somerset, where residents gathered to protest the Somerset Power Generating Station. Groups also convened in Holyoke and Salem Harbor.

THOUSANDS GATHER TO PROTEST COAL AND GLOBAL WARMING, WASHINGTON, D.C. On March 2, 2009, in the largest U.S. protest to date against global warming, several thousand demonstrators convened outside the Capitol Power Plant, calling on Congress to pass legislation to reduce greenhouse gases. Around 2,500 people blockaded the gates to the plant. No arrests were made. Just days before the planned protest, Speaker of the House Nancy Pelosi and Senate Majority Leader Harry Reid announced that the plant would be taken off coal and switched to natural gas. Many viewed the announcement as a victory for grassroots activism, but the rally went forward to call attention to coal issues around the country. Also on March 2, organizers of the Power Shift 2009 conference spearheaded a grassroots lobbying drive described as "the biggest lobbying day on climate and energy" in the history of the United States, with approximately four thousand students visiting almost every congressional office.

UNITED MOUNTAIN DEFENSE VOLUNTEER ARRESTED BY TVA. On March 4, 2009, United Mountain Defense volunteer staff person Matt Landon was arrested while driving a blind grandmother home after a public meeting through an unstaffed illegal TVA roadblock following the TVA Kingston Fossil Plant coal ash spill on December 22, 2008.

ACTIVISTS PROTEST MOUNTAINTOP REMOVAL, PETTUS, WEST VIRGINIA. On March 5, 2009, five activists were arrested for protesting at Massey Energy's Edwight mountaintop removal mine on Cherry Pond Mountain, unfurling a banner that read "Stop the blasting. Save the kids." The protesters were calling attention to the blasting taking place near a dam that holds 2.8 billion gallons of sludge and lies just a few hundred yards above the Marsh Fork Elementary School. All five were arrested.

"FREEZE ON COAL" AT MIDDLEBURY COLLEGE, VERMONT. On March 10, 2009, following the lead established by students at Santa Clara University, who convinced the school's president to divest the university from Massey Energy stock, forty students froze in place while getting lunch in the busiest cafeteria on campus. The activists held pieces of charcoal in their hands. The "freeze" lasted for two minutes, after which the students continued with their meal, explaining to onlookers what had just happened.

COUNCIL BUILDING BLOCKADE IN BRUSSELS, BELGIUM. On March 10, 2009, more than three hundred Greenpeace protesters blocked the entrances of the Council Building in Brussels to urge finance ministers to fix the climate. Protesters from twenty countries locked themselves to gates and fences while large contingents of anti-riot police and European Union security forces detained and arrested participants and secured the entrances.

PROTESTERS MARCH AGAINST COAL IN PALM SPRINGS, CALIFORNIA. On March 14, 2009, more than fifty people marched through downtown Palm Springs to call attention to the need for a moratorium on the construction of new coal-fired power plants. Protesters carried signs reading "Quit Coal Now!" The march was part of the Power Past Coal campaign, a hundred-day national action running from January 21 to April 30.

FOURTEEN ARRESTED AT TVA HEADQUARTERS IN KNOXVILLE, TENNESSEE. On March 14, 2009, local residents joined dozens of activists from across the country in a demonstration at the Tennessee Valley Authority headquarters. Police arrested fourteen individuals who staged a "die-in" in front of the building. This event was held in solidarity with communities affected by the destructive impacts of mountaintop removal coal mining and the survivors of the coal ash disaster in Harriman, Tennessee. The demonstration began with a rally in Market Square, where organizers from United Mountain Defense and Mountain Justice spoke about coal's impact from cradle to grave on communities in Appalachia and the surrounding area. The crowd then marched through downtown Knoxville and ended at TVA headquarters. At the end of the march, those participating in civil disobedience gave a statement about why they wanted to take this action. With the support of a singing crowd each participant fell to the ground, representing the deaths caused by the coal industry. After a few minutes Knoxville law enforcement informed participants that they were blocking the sidewalk and that they needed to remove themselves from the area. All fourteen people were arrested and cited for loitering.

ANTI-COAL PROTESTERS GATHER OUTSIDE STATEHOUSE IN TOPEKA, KANSAS. On March 19, 2009, over two hundred Kansas residents rallied on statehouse grounds to protest legislation that would resurrect two coal plants proposed for western Kansas. The group included environmentalists opposed to coal, steelworkers pushing to build wind turbines, rural advocacy groups, and Christian clergy. Bill 2182 would strip the Department of Health and the Environment of its power to regulate industry based on air quality concerns. The bill was vetoed by the governor.

"BLUEGRASS AT THE BANK" HITS BANK OF AMERICA BRANCH IN SARASOTA, FLORIDA. On March 20, 2009, Mountain Justice members and Earth First! activists from Florida and Appalachia disrupted the lobby of a Bank of America branch in Sarasota to protest the bank's continued funding of mountaintop removal mining and the construction of new coal-fired power plants. While several protesters distributed informational handouts about Bank of America's investments in coal to tellers and account holders, one individual played bluegrass banjo to celebrate the culture of the Appalachian region that the bank's investments threaten. The activists' signs read "Bank of America: still funding coal, killing communities."

PROTESTERS BLOCKADE COAL TERMINAL IN NEWCASTLE, AUSTRALIA. On March 21, 2009, hundreds of activists shut down the world's largest coal terminal to send a message to Australia to stop exporting coal. The blockade prevented coal carriers from entering Newcastle. The protesters paddled kayaks and boats made from milk crates and inner tubes.

RISING TIDE DISRUPTS COAL-TO-LIQUIDS CONFERENCE IN WASHINGTON, D.C. On March 26, 2009, activists with DC Rising Tide interrupted an industry conference to denounce coal-to-liquids technologies. The protesters stood in the audience and gave loud speeches refuting the statements of executives from Chevron, CONSOL Energy, the World Coal Institute, and the World Petroleum Council. Displaying banners including "Coal kills" and "Renewable energy now," activists called for an end to fossil fuels and for adoption of clean, renewable energy sources.

STUDENTS RALLY OUTSIDE CAPITOL IN AUSTIN, TEXAS. On March 30, 2009, student activists from ReEnergize Texas gathered at the capitol to rally for clean energy projects and green jobs. Members of the state legislature were also in attendance. The group expressed support for proposed legislation that would enact a temporary moratorium on coal plants without carbon capture and sequestration. After the rally, activists visited seventy-five legislative offices to lobby for the bill.

INANIMATE ACTIVIST WITH MANNEQUINS FOR CLIMATE JUSTICE SHUTS DOWN BANK OF AMERICA BRANCH IN BOSTON, MASSACHUSETTS. On March 31, 2009, a member of Mannequins for Climate Justice was found chained to the doors of the Kenmore Square Bank of America, preventing the bank from opening. Pinned to the protester was the following note: "Even a dummy like me can see that Bank of America's massive loans to coal companies and support for the epidemic of foreclosures and evictions have to stop now."

APRIL **GREENPEACE ACTIVISTS HOLD A "COAL CIRCUS" ON BOSTON COMMON.** On April 1, 2009, as part of the global Fossil Fools Day campaign, about twenty Greenpeace activists staged a "coal circus" to refute the coal industry's claims that coal plants can produce energy without significant greenhouse gas emissions. Protesters wore clown suits and put up a banner that read "The Coal Circus. It's So Clean! (April Fools)."

OVER A HUNDRED ARRESTED FOR ALLEGEDLY PLANNING DIRECT ACTION AGAINST COAL PLANT IN NOTTINGHAM, UNITED KINGDOM. On April 14, 2009, police carried out what may be the largest preemptive strike on environmental activism in British history, arresting 114 for allegedly planning a direct action at E.ON's Ratcliff-on-Soar plant. Caroline Lucas, leader of the Green Party, said, "Confidence in policing of protests like this has just about hit rock bottom. Peaceful protest is a civil liberty we need to uphold, even more in the context of the lack of government action on climate change. We have tried all the usual channels." The activists were charged with conspiracy to commit criminal damage and aggravated trespass.

ACTIVISTS ARRESTED AT MASSEY ENERGY MINE IN WEST VIRGINIA. On April 16, 2009, five people were arrested when activists from Climate Ground Zero unfurled a 40-foot-tall banner reading "EPA stop MTR" (mountaintop removal) at Massey Energy's Edwight mountaintop removal site. Massey had recently started blasting at the mine directly above the town of Naoma. Activists are concerned because the blasting is near a slurry dam, which poses a risk to the local Marsh Fork Elementary School.

HUNDREDS PROTEST IN CHARLOTTE, NORTH CAROLINA, AGAINST DUKE'S PROPOSED CLIFFSIDE PLANT. On April 20, 2009, hundreds of people marched and rallied against Cliffside in Charlotte, North Carolina. More than a dozen environmental, faith-based, and social justice groups organized the demonstration. Speakers called on Duke Energy and the state of North Carolina to cancel construction of the Cliffside plant. Forty-four activists were arrested.

ACTIVISTS BEGIN FAST TO URGE IMMEDIATE ACTION ON GLOBAL WARMING. On April 20, 2009, more than two hundred people from thirty states and six countries began fasting for up to forty days, in order to call attention to the need for the United States to demonstrate world leadership on climate change. The Fast For Our Future action called for legislation mandating a 25–40 percent or higher decrease in greenhouse gas emissions over 1990 levels, a moratorium on building new coal-fired power plants, and strong climate legislation containing no giveaways to polluters.

GREENPEACE ACTIVISTS HANG BANNER AT INTERNATIONAL CLIMATE MEETING IN WASHINGTON, D.C. On April 27, 2009, activists from Greenpeace USA hung a huge banner from a crane across the street from the State Department to urge action from ministers of the seventeen largest greenhouse gas emitters. The ministers were in D.C. to discuss climate change as part of the Major Economies Forum. The banner read "Too Big to Fail: Stop Global Warming—Rescue the Planet." Seven activists were arrested

MAY **ACTIVISTS PROTEST CLIFFSIDE PLANT AT DUKE ENERGY SHAREHOLDER MEETING.** On May 7, 2009, activists dominated Duke Energy's annual shareholder meeting in Charlotte, North Carolina. About twenty-five protesters gathered outside the company's headquarters, calling for Duke to cancel its proposed Cliffside Plant. Inside the meeting, activists owning shares in the company grilled CEO Jim Rogers about Duke's coal and nuclear investments.

SEVEN ARRESTED AT MASSEY ENERGY COMPLEX IN WEST VIRGINIA. On May 23, 2009, more than seventy-five residents of the Coal River Valley and members of a coalition that includes Mountain Justice and Climate Ground Zero picketed the entrance to Massey Energy's Marfork mining complex. The actions were in protest of the company's plans to blast 100 feet away from the Brushy Fork coal sludge impoundment. The demonstration began with a prayer and sermon by Bob "Sage" Russo of Christians for the Mountains. Referencing the Sermon on the Mount, he called upon citizens to be stewards of the Earth and to move towards sustainable, stable jobs. Protestors stood in front of the gates of the mine facility with signs including "7 billion spilled, 998 killed." "Passersby on Route 3 were overwhelming supportive with honks, waves, and thumbs up signs," Rock Creek (Raleigh County) resident Julia Sendor said. During the protest, seven people approached the entrance to the dam facility and the Whitesville detachment of the West Virginia State Police asked them to leave. When the seven refused, the State Police arrested them. After the arrests, former U.S. Congressman Ken Hechler, a longtime opponent of strip mining, gave a speech. He underscored the responsibility of citizens to safeguard their freedoms and stand up for their rights. The protest came just hours after activists carried out two non-violent direct actions to protest mountaintop removal and coal sludge impoundments. state police arrested eleven activists at two civil disobedience actions in West Virginia. In one action, six people locked themselves to mining equipment at a Patriot Coal mine on Kayford Mountain. Another group raised a 20-by-60-foot banner at Massey Energy's Brushy Fork coal slurry impoundment near Pettus. Protesters were part of a coalition that included Mountain Justice, Climate Ground Zero, and concerned citizens.

TWO PROTESTERS IN BOATS ARRESTED ON BRUSHY FORK IMPOUND-MENT, WEST VIRGINIA. On May 23, 2009, two protesters wearing hazmat suits and respirators were arrested after boating onto the Brushy Fork impoundment and floating a banner that read, "No More Toxic Sludge." State Police charged the activists with littering and misdemeanor trespass and transported them to the Southern Regional Jail. Bail has been set at $2,000.

POLICE REMOVE SIX ACTIVISTS FROM MOUNTAINTOP REMOVAL EQUIPMENT IN WEST VIRGINIA. Six people raised a "Never Again" banner and locked themselves to mining equipment at Massey Energy's Patriot Coal mine on Kayford Mountain. State Police arrived on site to find three people chained to the main axle of the truck and three others chained outside the truck's cab. The police removed the six activists, who, along with two others supporting them, were transported to the Madison County Courthouse, where they were reportedly processed and released. The protesters are part of a coalition that includes Mountain Justice, Climate Ground Zero, and concerned citizens.

4,000 PROTEST PROPOSED COAL PLANT IN MAINZ, GERMANY. On May 23, 2009, thousands of activists gathered in Mainz to protest an 820-megawatt coal-fired power plant being built on the banks of the Rhine river. The protesters carried banners and marched through the city to display their opposition to the new plant, which is expected to be operational by 2013. The group included local farmers, environmental activists, residents, students, and politicians.

JUNE

ACTIVISTS SCALE 20-STORY DRAGLINE AT MOUNTAINTOP REMOVAL SITE IN TWILIGHT, WEST VIRGINIA. On June 19, 2009, fourteen protesters visited the Massey Energy Twilight mountaintop removal site in Boone County, West Virginia, and climbed a twenty-story strip mining machine called a dragline. The activists unfurled a 15-by-150-foot banner reading, "Stop Mountaintop Removal. Clean Energy Now!" All fourteen protesters were arrested.

ACTIVISTS BOARD COAL SHIP IN KENT, ENGLAND. On June 21, 2009, ten Greenpeace activists boarded a ship delivering coal to the Kingsnorth Power Station. The group used inflatable speedboats to target the boat as it sailed up the River Medway. All ten protesters were arrested and charged with conspiring to commit criminal damage and having an unauthorized presence on a ship.

DOZENS ARRESTED PROTESTING AT MASSEY ENERGY SITE IN COAL RIVER VALLEY, WEST VIRGINIA. On June 23, 2009, 29 protesters including 94-year-old former United States congressman Ken Hechler, NASA climate scientist James Hansen, Goldman Prize Award winner Judy Bonds, Rainforest Action Network director Michael Brune, and actress Daryl Hannah were arrested at the entrance to a Massey Energy coal processing plant near the Marsh Fork Elementary School in Sundial. After being blocked from entering the facility by a crowd of Massey employees, the protesters sat down on state highway 3 and were arrested. Massey employees on scene behaved aggressively, heckling speakers at a rally preced ing the march to the plant entrance. One woman was arrested and charged with battery after striking Judy Bonds, co-director of Coal River Mountain Watch.

MORE THAN 700 PEOPLE TURN OUT AGAINST CARBON SEQUES-TRATION PROJECT IN GREENVILLE, OH. On June 29, 2009, more than 700 people attended a meeting organized by opponents of a proposed $92.8 million carbon capture and storage project in Ohio. The project would inject carbon dioxide from a nearby ethanol plant more than 3,000 feet underground. The group included local residents, activists, and politicians. A representative of the Ohio Environmental Council commented that he had "rarely seen a community that well organized and that strong."

BANNER DROP AT EPA HEADQUARTERS, BOSTON, MA. On June 29, 2009, activists with Rising Tide draped a 25-foot banner reading, "Mountain Top Removal Kills Communities: EPA No New Permits. MountainJustice.org" at the downtown offices of the Environmental Protection Agency. The group is urging the agency to block over 150 pending permits for mountaintop removal coal mining in West Virginia, Kentucky, and Virginia.

JULY   **GREENPEACE ACTIVISTS CLIMB MOUNT RUSHMORE.** On July 8, 2009, several Greenpeace activists climbed Mount Rushmore in South Dakota to hang a banner calling for action on climate change. The banner, which was sixty-five feet high by thirty-five feet wide, featured a portrait of President Obama and read, "America Honors Leaders Not Politicians: Stop Global Warming." The action was part of an effort to send a message to world leaders at the G8 meeting in L'Aquila, Italy.

GREENPEACE ACTIVISTS SPRAY-PAINT COAL SHIP AND POWER STATION IN ITALY. On July 10, 2009, a group of Greenpeace activists spray-painted the message "G8: Failed" on a ship carrying 25,000 tons of coal bound for the Civitavecchia power station near Rome. Farther south, activists in Brindisi painted "Stupid" on Italy's largest coal plant. The actions were intended to protest the G8 meeting on climate change, which UN official Yvo de Boer described as "disappointing."

MORE THAN 200 PEOPLE TURN OUT FOR RALLY AGAINST COAL PLANT IN BOULDER, CO. On July 14, 2009, more than 200 Boulder residents attended a rally opposing the Valmont Station. Activists from Greenpeace and Clean Energy Action planned the rally to draw attention to a hearing on renewing Valmont's air permit. Many Boulder residents are pushing for the plant to stop burning coal and switch to cleaner energy.

HUNDREDS RALLY FOR CLEAN ENERGY IN LANSING, MI. On July 29, 2009, hundreds of individuals gathered at the Michigan State Capitol to rally for the development of wind, solar, and other renewable energy sources instead of building new coal plant projects in the state. The rally was hosted by a coalition of environmental groups, including Clean Water Action, Michigan Interfaith Power and Light, the Michigan Land Use Institute, and the Sierra Club. Organizers said they hoped to convince lawmakers to expand investments in the state that support clean, renewable energy.

PROTESTERS BLOCK HAY PT. COAL TERMINAL IN AUSTRALIA. On August 5, 2009, Greenpeace activists used the group's largest ship to block BHP Billiton's coal terminal on the northwest coast of Australia. The action halted loading and shipments for more than 36 hours.

ACTIVISTS DUMP COAL OUTSIDE SOUTH LANARKSHIRE COUNCIL HEADQUARTERS IN HAMILTON, UK. On August 10, 2009, activists protesting plans for a new mine near Douglas, UK, dumped piles of coal outside the headquarters of South Lanarkshire Council. A damaged conveyor belt, which was suspected to be another action by climate change protesters, disrupted coal deliveries at an existing mine in the area. The protesters, organized by the Camp for Climate Action Scotland, said they wanted to call attention to the environmental and health issues of open cast mining.

ACTIVISTS LOCK DOWN WEST VIRGINIA DEPARTMENT OF ENVIRONMENTAL PROTECTION IN CHARLESTON, WV. On August 11, 2009, four protesters locked themselves to the entrance at the West Virginia DEP, displaying signs that read, "Closed Due to Incompetence" and "Department of Encouraging Pollution." The activists demanded that the EPA and the Office of Surface Mining, Reclamation, and Enforcement take over of the agency's programs. They also called for Secretary Randy Huffman's resignation.

"GOING AWAY PARTY" FOR NATIONAL COAL CORPORATION IN KNOXVILLE, TV. On August 13, 2009, an employee with National Coal Corporation forcefully removed a non-violent anti-mountaintop removal protester from the National Coal headquarters in West Knoxville. The protester was part of a group participating in "Love and Hug National Coal Month," part of a series of protests organized by United Mountain Defense every Thursday in August at National Coal's office. The protesters had organized a "Going Away Party" for NCC after the coal company defaulted on $60 million of loans in Alabama in July 2009. To mark this event the protestors brought balloons and cupcakes reading "Bye National Coal' and "Take a Hike." Wearing party hats and dancing to festive music, the volunteers entered the National Coal Headquarters in order to deliver the cupcakes. Within 30 seconds an employee of National Coal Corporation wrapped his hand around the video camera, contorted the cameraman's wrist, and escorted the peaceful group back outside, at which point he stated that NCC did not want to call the police. The non-violent protesters complied with the National Coal employee's request and moved to the public right of way in front of the office building. They educated passing motorists, gave away the unwanted cupcakes, danced, and had a fun time in the hot sun.

ACTIVISTS OCCUPY TREES TO STOP BLASTING IN COAL RIVER VALLEY, WV. From August 25 to 31, 2009, protesters from Climate Ground Zero and Mountain Justice occupied treetops at the edge of Massey Energy's Edwight mountaintop removal site in Raleigh County, West Virginia. The activists unrolled banners reading "Stop Mountain Top Removal" and "DEP: Don't Expect Protection." They were less than 30 feet from the mine and less than 300 feet from the blasting activity, which was forced to stop because of their close proximity. On the sixth day of the protest, the last activist finally descended and was arrested. A spokesman for Climate Ground Zero said sleep deprivation had been endangering the protesters.

■

APPENDIX B

Coal Plants Cancelled, Abandoned, or Put on Hold

■

2007

JANUARY   **HUNTER UNIT 4.** The Oregon Public Utility Commission rules that PacifiCorp had failed to prove a need for Hunter Unit 4, a proposed 575-megawatt coal plant in Castle Dale, Utah.

FEBRUARY   **BIG BROWN 3, MORGAN CREEK 7, TRADINGHOUSE 3 AND 4, SANDOW 5, MONTICELLO 4, MARTIN LAKE 4, LAKE CREEK 3.** As part of a buyout of Texas utility TXU by private equity firms, TXU abandons plans for eight out of eleven proposed plants in the state.

           CLIFFSIDE SECOND UNIT. The North Carolina Utilities Commission rejects one of the two 800-megawatt units at Duke Energy's Cliffside Steam Station Modernization proposal, citing increased construction costs. Opponents continue to fight the second unit.

MARCH   **CORN BELT PLANT.** Corn Belt Energy Corporation abandons plans to build a 91-megawatt coal plant in Illinois. The plant would have been financed by a grant from the United States Department of Energy.

MAY   **INDIAN RIVER POWER PLANT.** The Delaware Public Service Commission rejects NRG Energy's proposal for a 600-megawatt coal plant at its existing Indian River Power Plant in favor of an alternative wind/gas proposal.

ESCANABA PLANT. Wisconsin Public Power Inc. and the city of Escanaba, Michigan, cancel plans to build a 300-megawatt coal plant in Escanaba.

PACIFICORP PLANTS. Newest revision of Integrated Resource Plan provided by PacifiCorp to Oregon regulators omits four coal plants (locations not specified) that had been listed in previous plans.

JUNE

NUECES IGCC PLANT. Tondu Corp abandons plans for the Nueces IGCC plant in Corpus Christi, Texas, citing rising costs and uncertain construction schedules for IGCC. The company plans to build a gas plant instead.

JULY

TAYLOR ENERGY CENTER. Florida Municipal Power Agency withdraws its state permit application for the 800-megawatt Taylor Energy Center shortly after the Florida Public Service Commission rejects the Glades Power Plant.

GLADES POWER PLANT. The Florida Public Service Commission rejects the permit application of Florida Power & Light's 1960-megawatt Glades Power Plant citing, in part, uncertainty over the cost of future carbon regulations.

SALLISAW ELECTRIC GENERATING PLANT. Tenaska cancels its 660- to 880-megawatt Sallisaw Electric Generating Plant in Oklahoma on the grounds that it is not economically viable.

LS POWER SUSSEX PROPOSAL. The Sierra Club reports that LS Power and Dynegy have quietly abandoned plans for a 1600-megawatt coal plant in Sussex County, Virginia. The companies no longer list the plant on their websites.

AUGUST

THOROUGHBRED GENERATING STATION. Franklin Circuit Court reverses the air permit for Peabody Coal Company's 1500-megawatt Thoroughbred Generating Station in Kentucky due to inadequate air pollution control analysis.

SEMINOLE 3 GENERATING STATION. Florida's Department of Environmental Protection rejects the Seminole Electric Power Cooperative's 750-megawatt Seminole 3 Generating Station on the grounds that the plant would not minimize environmental and public health impacts, and would not serve the public interest.

NELSON CREEK PROJECT. Great Northern Power hasn't submitted an air permit application for its proposed 500-megawatt Nelson Creek coal plant in Circle, Montana. A GNP lobbyist testifies in a state legislative session that the company is no longer pursuing the project.

SOUTH HEART POWER PROJECT. Great Northern Power withdraws its air permit application for the 500-megawatt South Heart Power Project in North Dakota.

MESABA ENERGY PROJECT. The Minnesota Public Utilities Commission decides that Excelsior Energy's 600-megawatt Mesaba IGCC plant "is not in the public interest."

SEPTEMBER **HOLCOMB UNIT 3.** Sunflower Electric Power Cooperative's proposal for the 700-megawatt Holcomb Unit 3 is canceled after Colorado adopted a law requiring that rural electric cooperatives get 10 percent of their power from renewable resources.

RUSSELL STATION II. Rochester Gas and Electric, a subsidiary of Energy East, changes plans for the proposed 300-megawatt Russell Station II plant from coal to natural gas. The decision is based partly on public opposition to coal.

GASCOYNE 175 PROJECT. Westmoreland and Montana Dakota Utilities fail to begin construction of the North Dakota Gascoyne 175-megawatt power plant or request an extension of the air permit. As a result, the air permit is rendered invalid and the company must go through the air permitting process again if it intends to construct the plant.

ROUNDUP POWER PROJECT. Montana regulators revoke the air permit for Bull Mountain Development's 300-megawatt Roundup Power Project.

RED ROCK GENERATING STATION. American Electric Power and Oklahoma Gas & Electric's 950-megawatt Red Rock Generating Station is rejected by the Oklahoma Corporation Commission for failure to evaluate alternatives such as natural gas.

AVISTA PLANT. Avista Utilities in Washington completes its 2007 Integrated Resource Plan, eliminating at least one coal plant from its resource strategy. The company cites Washington state law prohibiting new coal plants whose emissions would exceed those of a natural gas plant.

BOWIE IGCC POWER STATION. Southwestern Power Group cancels its proposed 600-megawatt IGCC Bowie Power Station in Arizona in favor of pursuing a natural gas–fired plant. The company cites economics and regulatory uncertainty.

OCTOBER **HOLCOMB UNITS 1 AND 2.** Sunflower Electric Power Cooperative's proposal for the 1400-megawatt Holcomb Units 1 and 2 is denied an air permit by the Kansas Department of Health and

Environment (KDHE) due to concerns about global warming. The Director of KDHE states that it would be "irresponsible" to ignore global warming concerns when evaluating whether to build a new plant.

MARION GASIFICATION PLANT. Madison Power's 600-megawatt Marion Gasification Plant (IGCC) plant in Marion, Illinois has been placed on hold due to construction of a nearby supercritical coal plant which has hindered power demand and tied up transmission and coal transport infrastructure.

HUNTLEY GENERATING STATION. NRG's Huntley Generating Station (680-megawatt IGCC) is "on hold" because it "must find cost reductions to maintain state-awarded financial support."

BUFFALO ENERGY PROJECT. Buffalo Energy Partners IGCC plant in Wyoming has been canceled due to transmission constraints, rising construction costs, limited available technology guarantees and an unsuccessful bid for funding.

XCEL IGCC PLANT. Minneapolis-based Xcel Energy shelves plans for a 600-megawatt IGCC plant in Colorado for at least two years, citing rising construction costs and slowing demand.

POLK POWER STATION. Tampa Electric suspends a 630-megawatt expansion at its Polk Power Station. The decision is influenced by Florida Governor Charlie Crist's push to reduce carbon dioxide emissions.

WEST DEPTFORD PROJECT. Dynegy and LS Power cancel a proposed 500-megawatt coal plant in West Deptford, New Jersey, proposing instead a natural gas plant.

NOVEMBER **STANTON ENERGY CENTER.** Two months after breaking ground, Orlando Utilities Commission and Southern Company shelve plans for the 285-megawatt Stanton Energy Center, an IGCC plant in Orange County, citing concerns about future carbon controls in Florida.

PACIFIC MOUNTAIN ENERGY CENTER. Energy Northwest's application for the 793-megawatt Pacific Mountain Energy Center in Kalama, Washington, is suspended by state regulators because of insufficient plans for carbon sequestration.

TWIN RIVER ENERGY CENTER. Voters in Wiscasset, Maine, defeat two ballot measures that would have allowed a variance from local height limits, effectively blocking Point East from pursuing its proposed 700-megawatt coal plant, Twin River Energy Center.

MATANUSKA POWER PLANT. Matanuska Electric Association cancels plans to build a 100-megawatt coal plant in the Matanuska-Susitna Borough of Alaska. Local opposition by elected officials and increased costs are cited as reasons.

IDAHO POWER PROJECT. Idaho Power Company cancels plans to produce 250 megawatts from coal-fired plants by 2013. Instead, the company adopts new plans to add 101 megawatts of wind power and 45.5 megawatts of geothermal power by 2011, and to develop a natural gas turbine in Idaho by 2012.

DECEMBER   **ELMWOOD ENERGY CENTER.** Indeck Energy Services declines to renew the option for the property the company intended to use for the 660-megawatt Elmwood Energy Center in Elmwood, Illinois, indicating that it did not intend to pursue the project further. In September 2006 the U.S. EPA's Environmental Appeals Board had overturned the air permit, finding that it lacked emissions control requirements and environmental impact assessments.

RENTECH ENERGY MIDWEST. Rentech puts coal-to-liquids plant slated for East Dubuque, Illinois on indefinite hold, citing "pressure" put on the project by a lack of national CO_2 policy.

ALCOA PROJECT. Alcoa scraps plan to build a 950-megawatt coal plant at the site of a shuttered aluminum smelter in Frederick County, Maryland.

JIM BRIDGER STATION EXPANSION. Idaho Power and PacifiCorp abandon plans for a 600-megawatt expansion of the Jim Bridger Station, a power plant jointly owned by the two companies in Wyoming. A spokesman for PacifiCorp cites the uncertainty around coal, and states the company is looking at natural gas and wind power projects instead.

SOUTHERN ILLINOIS CLEAN ENERGY CENTER. Steelhead Energy's 545-megawatt IGCC proposal, the Southern Illinois Clean Energy Center, is declared inactive by the EPA.

SODA SPRINGS PROJECT. Mountain Island Energy abandons plans for a 600-megawatt coal plant in Soda Springs, Idaho, which had been announced in January 2007.

JIM BRIDGER IGCC DEMONSTRATION PROJECT. PacifiCorp and the state of Wyoming cancel their jointly sponsored IGCC and coal sequestration demonstration project at the Jim Bridger Station, which had been scheduled for operation in 2013.

INTERMOUNTAIN POWER PROJECT EXPANSION. PacifiCorp scraps plans for a 950-megawatt expansion of the Intermountain Power coal plant in Utah. The cancellation comes after six California cities that rely on the plant refused to support the expansion in July 2007; two other cities refused power contracts with the plant earlier in the year.

KANSAS CITY BOARD OF PUBLIC UTILITIES PROJECT. Following the denial of permits for Sunflower's Holcomb plants, the Kansas City Board of Public Utilities abandons plans to build a 235-megawatt coal plant at Nearman Creek in Wyandotte County.

WESTAR ENERGY KANSAS PLANT. Having announced that it was placing siting plans for a new 600-megawatt coal plant on hold due to rapidly escalating costs, Westar Energy, Kansas's largest utility, pursues regulatory approval for 295 megawatts of new wind capacity.

BETHEL POWER PLANT. After being on hold for over two years due to siting issues, the 100-megawatt Bethel Power Plant is abandoned by Nuvista Light and Power. (Month unknown.)

ROSEMOUNT PROJECT. Xcel Energy abandons plans to build a 550-megawatt coal plant near Rosemount, Minnesota. (Month unknown.)

RAY D. NIXON POWER PLANT EXPANSION. Having lost its partner on the project (Foster-Wheeler, which was delisted from the New York Stock Exchange), Colorado Springs Utilities abandons its 150-megawatt Ray D. Nixon Power Plant expansion proposal. (Month unknown.)

FAYETTE COUNTY ECONOMIC DEVELOPMENT PROJECT. Clean Coal Power Resources abandons the Fayette County Economic Development Project, a synthetic fuels project in Illinois. (Month unknown.)

BALDWIN ENERGY COMPLEX. Dynegy abandons its proposed 1300-megawatt Baldwin Energy Complex in Baldwin, Illinois. (Month unknown.)

ILLINOIS ENERGY GROUP PROJECT. Illinois Energy Group abandons its 1500-megawatt project in Franklin County, Illinois. (Month unknown.)

ELKHART PROPOSAL (TURRIS). Turis Coal Company abandons a 25- to 35-megawatt coal plant proposal in Elkhart, Illinois. (Month unknown.)

2008

JANUARY **AES COLORADO POWER PROJECT.** AES withdraws its application with the Colorado Department of Public Health and Environment to build the AES Colorado Power Project, a 640-megawatt coal-fired power plant in Washington County, west of Akron, Colorado.

HIGH PLAINS ENERGY STATION. Dynegy and LS Power withdraw their application with the Colorado Department of Public Health and Environment for a permit to build the High Plains Energy Station, a 600-megawatt coal-fired power plant in Morgan County, Colorado.

BUICK COAL AND POWER PROJECT. Radar Acquisitions Corp., of Calgary, Alberta, announces that its deal with West Hawk Development to explore the possibility of an IGCC plant near Limon, Colorado, where it owns 22,500 acres of surface rights and coal, has been terminated, putting an end to the proposed Buick Coal and Power Project.

FUTUREGEN. The U.S. Department of Energy cancels plans to build the experimental FutureGen plant in Mattoon, Illinois, which would attempt to capture and store its carbon dioxide emissions.

FEBRUARY **BIG CAJUN II UNIT 4.** In an NRG conference call, Robert Flexon, NRG's chief financial officer, states that the company has abandoned the Big Cajun II Unit 4 project in Louisiana, due to the fact that the company has only been able to contract out 450 of the plant's 705 megawatts.

MARCH **KENAI BLUE SKY PROJECT.** Agrium Corp. says a combination of rising construction costs and a worsening U.S. economy has convinced the company not to proceed with the Kenai Blue Sky Project, a coal gasification facility and adjacent electrical generating plant that the company had planned to build at its fertilizer plant on the Kenai Peninsula in Alaska.

NORBORNE BASELOAD PLANT. Associated Electric Cooperative announces that it is canceling its 660-megawatt Norborne Baseload Plant. The company cites three reasons: (1) an increase in costs to $2 billion due to "worldwide demand for engineering, skilled labor, equipment and materials"; (2) the Rural Utilities Service's cancellation of financing for coal projects; and (3) the increased regulatory and cost uncertainties surrounding carbon dioxide. The coop plans to pursue a combination of efficiency measures, wind power, and nuclear power.

APRIL   **MOUNTAINEER IGCC.** West Virginia's State Corporation Commission (SCC) rejects Appalachian Power Company's Mountaineer IGCC plant, a proposed 629-megawatt facility in Mason County. According to the SCC, Appalachian Power's estimate of $2.33 billion, which has not been revised since November 2006, is "not credible."

        SIU POWER PLANT. Southern Illinois University announces that it is putting plans to build a new IGCC plant on hold due to financial reasons after a feasibility study placed the cost of the 200-megawatt facility at $1.5 billion.

MAY   **GASCOYNE 500 PROJECT.** Westmoreland Power announces that it is suspending development of the Gascoyne 500 Project in North Dakota and returning $562,500 in state state subsidies for the project. The company cites lack of a customer for the power and uncertainty over carbon regulation. Company spokesman Keith Alessi writes to the N.D. Industrial Commission: "There is much uncertainty in the utility sector on when future carbon regulation will come into effect. This has slowed the development of coal-fired power plants.... At this time (we) cannot predict when a long-term customer (for the plant's electricity) can be found and the actual plant construction could commence."

JUNE   **MILTON YOUNG 3.** Minnkota Power Cooperative announces that it will delay building the Milton Young 3 station until 2026. In the meantime, the company has agreed to an arrangement with FPL Energy under which Minnkota will receive 99 megawatts of peak output from FPL's wind farm in Cavalier County, North Dakota

AUGUST   **GILBERTON COAL-TO-CLEAN-FUELS AND POWER PROJECT.** Coal magnate John Rich admits that because of ballooning costs and lack of government support he has given up on setting a timeline for the construction of the Gilberton Coal-to-Clean-Fuels and Power Project, the nation's first coal-to-oil plant, in Gilberton, Pennsylvania.

        SITHE SHADE TOWNSHIP PROJECT. Sithe Global Power halts plans to build a waste-coal plant in Shade Township, Pennsylvania due to lack of progress in finding a financing partner for the project. (The company says it will continue to seek permits.)

        LOWER COLUMBIA CLEAN ENERGY CENTER. Westward Energy fails to submit a siting application to the Oregon Energy Facility Siting Board, causing observers to conclude that the Lower

Columbia Clean Energy Center, a 520-megawatt IGCC plant, has been abandoned.

TWIN OAKS POWER UNIT 3. Albuquerque, New Mexico–based PNM says it will not pursue its 600-megawatt Twin Oaks Power Unit 3 expansion in Robertson County, Texas. In 2007 PNM signed a non-binding letter of intent to contribute the existing Twin Oaks station to EnergyCo, an unregulated energy joint venture with Bill Gates's Cascade Investments unit.

SEPTEMBER **WESTERN GREENBRIER CO-PRODUCTION DEMONSTRATION PROJECT.** Western Greenbrier Co-Production issues an official news release stating that it has canceled the Western Greenbrier Co-Production Demonstration Project, a 98-megawatt circulating fluidized bed coal plant in West Virginia. The United States DOE had previously notified company officials that it was pulling all funding for the project.

OCTOBER **BUFFALO ENERGY PROJECT.** The plan by Buffalo Managers and Montgomery Energy Partners to build the Buffalo Energy Project, a 1100-megawatt IGCC plant in Glenrock, Wyoming, appears to have been abandoned.

BENWOOD PLANT. CONSOL Energy and Synthesis Energy Systems cancel a large synthetic fuels plant in Benwood, West Virginia. The plant would have produced 720,000 metric tons of methanol and 100 million gallons of 87-octane gasoline per year using coal from CONSOL's Shoemaker Mine, with additional coal brought in from the McElroy and Loveridge mines, which also are owned by CONSOL. Synthesis cites hard economic times, unwillingness to commit equity capital, and a drop in oil prices. CONSOL expresses interest in continuing to pursue coal-to-liquids in the Northern Panhandle region of West Virginia, but says it will need a partner.

NOVEMBER **NELSON DEWEY GENERATING FACILITY EXPANSION.** Wisconsin state regulators vote unanimously to reject the Nelson Dewey Generating Facility expansion, citing concerns about global warming as well as the plant's $1.3-billion price tag, which has ballooned almost 60 percent from 2007 due to rising construction costs. The PSC notes that the likelihood of future regulations on carbon emissions will make it difficult for any new coal plant to be built in Wisconsin.

INDIANA SNG. Leucadia National Corporation, the main sponsor of the Indiana SNG project, requests the Utility Regulatory Commission to put its permitting decisions on hold due to difficulties securing commitments from potential buyers of the synthetic natural gas. If built, the project would have converted three million tons of coal annually to substitute natural gas.

KENTUCKY MOUNTAIN POWER. The Kentucky State Office of Adminstrative Hearings rejects Kentucky Mountain Power, a coal plant proposed in Calvert City (Knott County), Kentucky by EnviroPower. The company was previously granted an air permit from the state of Kentucky to build a 500-megawatt circulating fluidized bed coal plant.

DECEMBER   **LIMA ENERGY PLANT.** According to the Sierra Club, discussions with Global Energy officials have revealed that the Lima Energy plant in Ohio has been abandoned.

THOROUGHBRED. Peabody Energy withdraws its permit application to build two 750-megawatt coal-burning plants at its Thoroughbred campus in Kentucky.

2009

JANUARY   **ELK RUN ENERGY STATION.** LS Power announces that because of the economic downturn, it is canceling plans to build the Elk Run Energy Station in Waterloo, Iowa. A week before the cancellation, Dynegy agreed to dissolve its development venture with LS Power, in part because of the credit crisis.

HIGHWOOD GENERATING STATION. Developers of the Highwood Generating Station in Montana vote to halt work on the plant, citing regulatory uncertainty and environmental lawsuits. Instead developers will pursue a 120-megawatt plant that will be powered by natural gas with wind turbines for additional power.

MALMSTROM AIR FORCE BASE COAL-TO-LIQUIDS. Air Force officials announce that they have rejected construction proposals for the Malmstrom Air Force Base Coal-to-Liquids plant in Montana, and that they will no longer pursue development of the large synthetic fuel plant.

FEBRUARY   **AES SHADY POINT.** AES announces that it has withdrawn its air permit application for a new 650-megawatt unit at its Shady Point facility. Company spokesman Lindy Kiger explains the decision to cancel the project as "part of our broader strategy to re-evaluate our growth plans."

ELY ENERGY CENTER. Nevada Power announces that it is postponing its Ely Energy Center indefinitely because of increasing economic and environmental uncertainties. According to CEO Michael Yackira, the plant could be delayed for up to 10 years, or until carbon capture and storage technologies are available.

MARCH **LITTLE GYPSY REPOWERING PROJECT.** The Louisiana PSC orders Entergy Louisiana to suspend the Little Gypsy Repowering project, citing lower gas prices, escalating construction costs, and pending regulation of carbon by the Obama administration.

WHITE PINE ENERGY STATION. LS Power notifies Nevada state regulators that it is withdrawing its application to build the White Pine Energy Station, citing economic conditions and regulatory uncertainties. Instead, LS will focus on completing a planned 500-mile transmission line project to provide new access to renewable energy resources across Nevada.

SUTHERLAND GENERATING STATION UNIT 4. Alliant Energy subsidiary Interstate Power and Light Company announces that it is canceling the proposed Sutherland Generating Station Unit 4 in Iowa. The company says the decision was based on a combination of factors, including the financial climate and concerns about the possibility of future regulation of greenhouse gas emissions.

APRIL **UNNAMED TRI-STATE PLANT.** Because of the current economic climate and ongoing uncertainty in federal and state regulations, Tri-State Generation and Transmission announces that it will revisit its long-term resource plan, including options for new coal-fired power plants. An unnamed Tri-State coal plant earlier included in the National Energy Technology Laboratory report "Tracking New Coal-Fired Power Plants" is now on hold.

MAY **NEXTGEN ENERGY FACILITY.** Basin Electric Power Cooperative tells the South Dakota Department of Environment and Natural Resources that it is placing the NextGen Energy Facility on hold "because of the current regulatory, technology, and economic uncertainties."

MIDLAND POWER PLANT. Mid-Michigan Energy, a subsidiary of LS Power, announces that it is canceling the 750-megawatt Midland Power Plant in Michigan. The company cites "regulatory and economic uncertainty."

JUNE **NORTHERN MICHIGAN UNIVERSITY RIPLEY ADDITION.** Northern Michigan University announces that it is abandoning plans to build a 10-megawatt coal-fired power plant in favor of a wood-burning plant. The decision comes in the wake of a decision by the Environmental Appeals Board to remand the air permit for the project back to the Michigan Department of Environmental Quality due to failure to consider carbon dioxide emissions.

JULY **INTERMOUNTAIN POWER PROJECT UNIT 3.** Intermountain Power Agency officially cancels plans for the Intermountain Power Project Unit 3 expansion in Utah. The plant was initially canceled in July 2007, after six California cities that rely on the plant refused to support the expansion; two other cities refused power contracts with the plant earlier that year. The project was brought back to life when the Utah Associated Municipal Power System filed a lawsuit in January 2008 to force the Los Angeles Department of Water and Power to move forward with the third unit.

■

Notes

■

ONE

The 80% Solution

3 **Dana Milbank sounded mystified.** Dana Milbank, "Burned Up About the Other Fossil Fuel," *Washington Post*, June 24, 2008.

3 **Biographical background on James Hansen:** Mark Bowen, *Censoring Science: Inside the Political Attack on Dr. James Hansen and the Truth About Global Warming* (Dutton, 2007); Elizabeth Kolbert, *Field Notes From a Catastrophe: Man, Nature, and Climate Change* (Bloomsbury 2006); Elizabeth Kolbert, "The Catastrophist," *The New Yorker*, June 29, 2009.

5 **In 1981 he published a paper in *Science* predicting:** James Hansen et al., "Climate impact of increasing atmospheric carbon dioxide." *Science* 213 (1981), 957–966.

5 **Hansen's predictions:** Elizabeth Kolbert, "The Catastrophist," *The New Yorker*, June 29, 2009.

5 **In fact, the twelve-year period from 1997 to 2008 included the ten hottest years on record.** *Global Temperature Trends: 2008 Annual Summation*, Goddard Institute for Space Studies, updated January 13, 2009, data.giss.nasa.gov/gistemp/2008/.

5 **"The way I look at it, the great fun in science..."** Mark Bowen, *Censoring Science*, 274.

6 **"He's transparently full of integrity...."** Mark Bowen, *Censoring Science*, 71.

6 **"The work that he did..."** Elizabeth Kolbert, "The Catastrophist," *The New Yorker,* June 29, 2009.

6 **"I have a whole folder in my drawer labeled 'Canonical Papers.' About half of them are Jim's."** Elizabeth Kolbert, "The Catastrophist," *The New Yorker,* June 29, 2009.

7 **"It is unequivocal that the climate is changing..."** Joint Science Academies' Statement on Growth and Responsibility: Sustainability, Energy Efficiency and Climate Protection, accessed August 7, 2009.

7 **Going even further, he had led a coordinated research effort by an international assembly of climate scientists...** The result of this investigation is the paper James Hansen et al., 2008: "Target atmospheric CO_2: Where should humanity aim?" *Open Atmos. Sci. J.,* 2 (2008), 217–231.

7 **In discussing the results of this new research, Hansen was now back in the public arena...** Hansen's commentaries, articles, and papers can be found on his Columbia University website at columbia.edu/~jeh1/.

9 **Ending emissions from coal, he said, "is 80% of the solution to the global warming crisis."** Letter from James Hansen to Nevada governor Gibbons, April 14, 2008, columbia.edu/~jeh1/.

TWO
151 Time Bombs

15 **Erik Shuster's list:** *Tracking New Coal-Fired Power Plants* (National Energy Technology Laboratory, May 31, 2007) is archived at cmNOW.org. The most recent release of the report is posted at netl.doe.gov/coal/refshelf/ncp.pdf.

18 **Existing fleet of about 600 coal plants:** For aerial photos and data on each coal plant in the United States, see "Existing U.S. Coal Plants," CoalSwarm wiki, sourcewatch.org/index.php?title=Existing_U.S._Coal_Plants.

19 **Burn its way through a 125-car trainload of coal in two days.** This calculation is based on Wyoming sub-bituminous coal rated at 9,000 Btu/lb and a 31 percent plant efficiency rate.

19 **Creating a quantity of carbon dioxide that weighs approximately twice as much as the original train.** This calculation is based on Wyoming sub-bituminous coal rated at 9,000 Btu/lb and a 31 percent plant efficiency rate. Note that the amount of carbon dioxide produced by the trainload of coal would be even higher if coal were pure carbon, since the atomic weight of a carbon atom is 16 and the atomic weight of a carbon dioxide molecule is 44. However, coal is not entirely carbon and not all the carbon in coal is transformed into heat. Also, a small portion of the heat value of coal comes from oxidizing hydrogen into water. Overall, researchers have calculated that typical Wyoming sub-bituminous

coal produces 212.7 pounds of carbon dioxide per million Btu. See B.D. Hong and E.R. Slatick, "Carbon Dioxide Emission Factors for Coal," *Quarterly Coal Report,* January-April 1994, DOE/EIA-0121(94/Q1), eia/doe.gov/cneaf/coal/ quarterly/co2_article/co2.html.

19 **850,000 SUV drivers would have to switch to Priuses.** Comparison based on a 2009 model 4-wheel-drive Ford Explorer driven 12,000 miles/year (14,673 pounds of carbon dioxide) and a 2009 Toyota Prius driven 12,000 miles/year (4,995 pounds of carbon dioxide). Source: Terrapass Carbon Footprint Calculator, terrapass.com/carbon-footprint-calculator/.

19 **Background on coal.** Three recent books are Barbara Freese, *Coal: A Human History* (Penguin 2003), Jeff Goodell, *Big Coal: The Dirty Secret Behind America's Energy Future* (Houghton Mifflin 2006), and Richard Heinberg, *Blackout* (New Society Publishers, 2009).

20 **Cheney task force:** Michael Abramowitz and Steven Mufson, "Papers Detail Industry's Role in Cheney's Energy Report," *Washington Post,* July 18, 2007; Robert F. Kennedy, Jr., *Crimes Against Nature: How George W. Bush and His Corporate Pals Are Plundering the Country and Hijacking Our Democracy,* (HarperCollins, 2004), Chapter 7, "King Coal."

20 **More on the plans to expand U.S. coal capacity from the perspective of the coal industry:** *Opportunities to Expedite the Construction of New Coal-Based Power Plants,* National Coal Council, December 2004.

THREE
Inside the Swarm

This chapter is based on Ted Nace, "Stopping Coal in Its Tracks," *Orion Magazine,* January/February 2008. Adapted with permission of The Orion Society.

27 **As much as 88 percent of the coal's carbon dioxide can be captured in an IGCC plant, along with 99 percent of its sulfur oxides and particulates and 95 percent of its mercury:** *The Future of Coal,* Massachusetts Institute of Technology, 2007.

28 **"What we are exploring..."** Bob Burton, "Australia Pursues Greenhouse Gas Burial as Climate Solution," *Environment News Service,* March 1, 2004.

28 **According to a study by engineers at Massachusetts Institute of Technology:** *The Future of Coal,* Massachusetts Institute of Technology, 2007.

28 **The Department of Energy estimated that by the end of the century, the amount of liquified carbon dioxide needing to be permanently sequestered would be enough to fill Lake Erie twice over or cover the entire state of Utah with a blanket of liquified carbon dioxide 14 feet thick.** This calculation is based on an estimate of 6 billion tons per year of current U.S. carbon dioxide

emissions, and an estimate that cumulative emissions from 2004 to 2100 will be 1 trillion tons, under a reference case scenario. Assuming 68 pounds per cubit foot, such a scenario would produce 200 cubic miles of liquefied carbon dioxide. The volume of Lake Erie is 113 cubic miles. Source: Jay Braitsch, "Carbon Capture and Storage: DOE/Office of Fossil Energy Programs," Presentation at Edison Foundation CCS Conference, March 4, 2008, edisonfoundation.net/ events/2008-03-03/Braitsch_presentation.pdf.

29 **"The notion of coal as the solution to America's energy problems is a technological fantasy on par with the dream of a manned mission to Mars."** Jeff Goodell, "The Dirty Rock," *The Nation*, April 19, 2007.

29 **The cost of building such plants was expected to be around 40 percent higher:** See for example *The Future of Coal*, Massachusetts Institute of Technology, 2007, Table 3.5: "Representative Performance and Economics for Oxy-Fuel Pulverized Coal and IGCC Power Generation," which estimates construction costs for a conventional super-critical pulverized coal plant at $1,330 per kilowatt and for an IGCC plant with carbon capture and storage at $1,890 per kilowatt (a 42 percent increase).

29 **In all, each plant would have to burn about 25 percent more coal:** See for example *The Future of Coal*, Massachusetts Institute of Technology, 2007, Table 3.5: "Representative Performance and Economics for Oxy-Fuel Pulverized Coal and IGCC Power Generation," which estimates fuel costs for a conventional super-critical pulverized coal plant at $1.33 per kilowatt-hour and for an IGCC plant with carbon capture and storage at $1.64 per kilowatt-hour (a 23 percent increase).

31 **Subsidies for the Mesaba project:** *The Economics of the Mesaba Energy Project*, Citizens Against Mesaba Project, December 17, 2006, camp-site.info/resources. html#cdocs.

FOUR

But We'll Freeze in the Dark!

35 **Will Offensicht quote:** Will Offensicht, "Must We Freeze in the Dark?" Scragged. com, February 14, 2008.

36 **University of Delaware public opinion survey showed strong support for wind and strong opposition to increased coal generation:** Jeremy Firestone, Willett Kempton and Andrew Krueger, *Delaware Opinion on Offshore Wind Power: Interim Report* (January 16, 2007), University of Delaware College of Marine and Earth Studies. The study reported that 55.3% of Delaware residents agreed that wind farms in state oceanic waters should be "encouraged and promoted," 36.7% said they should be "allowed in appropriate circumstances," 3.1% said they should be "tolerated," and 0.7% said they should be "prohibited

in all instances." When asked about wind farms sited in Delaware Bay, support dipped slightly to 47.3% for "encouraged and promoted, 39.1% for "allowed in appropriate circumstances," 5.3% for "tolerated," and 2.7% for "prohibited in all instances." With respect to coal versus wind, the researchers reported: "91.1 percent of the responses would vote to expand electricity with offshore wind power rather than coal or natural gas, when told they would pay more for the wind power."

37 **Lazard study:** *Levelized Cost of Energy Analysis, version 2.0*, Lazard, June 2008. Results summarized at "Comparative Electrical Generating Costs," CoalSwarm wiki, sourcewatch.org/index.php?title=Comparative_electrical_generation_costs.

37 **California Energy Commission study:** *CPUC GHG Modeling,* Energy and Environmental Economics, Inc., at http://www.ethree.com/cpuc_ghg_model html. Accessed June 2008. For summary of results and details on accessing the spreadsheet containing cost components and assumptions, see "Comparative Electrical Generating Costs," CoalSwarm wiki, sourcewatch.org/index. php?title=Comparative_electrical_generation_costs.

37 **Department of Energy wind study:** *20% Wind Energy by 2030,* U.S. Department of Energy, May 12, 2008, 9–12.

38 **During 2007, numerous solar thermal plants were moving forward:** For a current list of projects, see "List of solar thermal power stations," Wikipedia, en.wikipedia.org/wiki/List_of_solar_thermal_power_stations.

39 **Estimates of land area for solar thermal plants:** "Concentrating solar power land use," CoalSwarm wiki, sourcewatch.org/index.php?title=Concentrating_solar_power_land_use; David R. Mills and Robert G. Morgan, "Solar Thermal Electricity as the Primary Replacement for Coal and Oil in U.S. Generation and Transportation," wired.com/images_blogs/wiredscience/files/MillsMorganUSGridSupplyCorrected.pdf.

39 **Comparison of land disturbed by solar thermal to land disturbed by coal mining:** The David Mills calculation for solar thermal land use amounts to 9,025 square miles. In contrast, the amount of land disturbed to date by coal mining operations has been estimated at 9,000 square miles, including 1,644 square miles disturbed by current operations. Source: *On the Rise: Solar Thermal Power and the Fight Against Global Warming,* Environment America, Spring 2008 (PDF file), p. 28. That study cites as the source for the 9,000-square-mile figure Adam Serchuk, *The Environmental Imperative for Renewable Energy: An Update,* Renewable Energy Policy Project, April 2000; as the source for the 1,644-square-mile figure the study cites U.S. Department of the Interior, Office of Surface Mining, "Answers to the 10 Most Frequently Asked Questions," osmre. gov/answers.htm, March 26, 2008.

40 **California and United States per capita electricity usage:** Anant Sudarshan and James Sweeney, "Deconstructing the 'Rosenfeld Curve'," Precourt Institute for Energy Efficiency Working Paper, July 1, 2008.

41 **Art Rosenfeld biographical information:** Arthur H. Rosenfeld, "The Art of Energy Efficiency: Protecting the Environment with Better Technology," *Annual Rev. Energy Environ.* 24 (1999), 33–82; "President Bush Names Arthur Rosenfeld the 2005 Enrico Fermi Award Winner," U.S. Department of Energy press release, April 27, 2006.

44 **Guy Caruso caused 132 coal plants to disappear:** Ted Nace, "The Magic Mouse of Guy Caruso," *Grist*, March 21, 2008.

45 **Google Clean Energy 2030:** Jeffery Greenblatt, *Clean Energy 2030: Google's proposal for reducing America's dependence on fossil fuels*, Google knoll, knol. google.com/k/jeffery-greenblatt/clean-energy-2030/15x31uzlqeo5n/1#.

45 **Eric Schmidt:** Alan Murray, "The Search for Change: Eric Schmidt of Google on why the company spends so much time worrying about energy," *Wall Street Journal*, March 9, 2009.

FIVE

What About China?

47 **According to a frequently cited statistic, the country was building the equivalent of two mid-sized coal plants each week, amounting to a yearly increase equivalent to the entire U.K. power grid each year:** See, for example, "King Coal," *National Geographic*, May 2008, 144.

47 **In 2008 annual coal output in China reached 2.76 billion metric tons:** *Coal Statistics 2008*, World Coal Institute, worldcoal.org/resources/coal-statistics/.

48 **an analysis of worldwide reserve figures completed in 2007:** *Coal: Resources and Future Production*, Energy Watch Group, EWG-Series No. 1/2007, March 2007.

49 **"[I]t is not possible to confirm the often-quoted suggestion that there is a sufficient supply of coal for the next 250 years":** National Reserach Council, *Coal: Research and Development to Support National Energy Policy*, 2007, 44–45.

49 **USGS study of the Gillette field:** James Luppens et al., *Assessment of Coal Geology, Resources, and Reserves in the Gillette Coalfield, Powder River Basin, Wyoming*, U.S. Geological Survey Open-File Report 2008-1202, 2008.

51 **six large wind farms:** Keith Bradsher, "Drawing Critics, China Seeks to Dominate in Renewable Energy," *New York Times*, July 14, 2009.

51 **China's contribution to the greenhouse gases currently in the atmosphere, a reflection of historical consumption, remains lower than that of the United States not only on a per capita basis but on an absolute basis:** For the period 1751–2006, China contributed 8.2% of cumulative emissions while the United States contributed 27.5% of cumulative emissions. Source: Letter from James

Hansen to Australian prime minister Kevin Rudd, March 27, 2008, Figure 4: "Annual and cumulative fossil fuel CO_2 emissions by country of emission," columbia.edu/~jeh1/.

SIX
Hurricane Politics

54 **In the Pacific Northwest, Washington governor Christine Gregoire signed a bill:** For details on the related emissions controls of Washington and California, see "Schwarzenegger clause," CoalSwarm wiki.

54 **British Columbia, New Zealand, and Ontario initiatives:** For details see "Coal moratorium," CoalSwarm wiki.

59 **A St. Petersburg Times poll released in May 2007:** Jennifer Liberto, "Poll: Most Floridians favor action on global warming," St. Petersburg Times, May 14, 2007.

59 **"Sometimes, the political system is like the climate system..."** Will Dana, "Gore 3.0," Rolling Stone, June 28, 2006.

60 **Meetings between Chris Kise and utility officials:** Steve Bousquet and Craig Pittman, "Fla. utilities dump coal fired power plant: Gov. Charlie Crist says climate change played a role in plans," St. Petersburg Times, July 4, 2007.

61 **Yale University and University of Miami poll:** Leiserowitz, A. and Broad, K., "Florida: Public opinion on climate change," A Yale University / University of Miami / Columbia University Poll (New Haven, CT: Yale Project on Climate Change), 2008.

SEVEN
Kansas

64 **The Edison Electric Institute, which represented large power utilities, wrote, "only in the imagination ... does there exist any widespread demand for electricity on the farm or any general willingness to pay for it."** Jack Doyle, Lines Across the Land: Rural Electric Cooperatives—The Changing Politics of Energy in Rural America, (Environmental Policy Institute, 1979), 4.

65 **In 1964 the rural electrics came under attack as communist institutions, and federal funding for the movement became an issue in the presidential campaign between Lyndon Johnson and Barry Goldwater:** Jack Doyle, Lines Across the Land, 10.

65 **Goldwater said that the Rural Electrification Administration had "outlived its usefulness."** Jack Doyle, Lines Across the Land, 10.

66 **"Within your organization you have a much more potent force at your fingertips..."** Jack Doyle, *Lines Across the Land*, 11.

66 **As time passed, what had once been among the most progressive organizations in America became in many ways one of the least:** For a current critique of the rural electric cooperatives, see Jim Cooper, "Electric Co-operatives: From New Deal to Bad Deal?" *Harvard Journal on Legislation*, 2008. For a broader account of the anti-environmental positions taken by much of the rural electric establishment in response to the major environmental legislation of the 1960s and 1970s, see Jack Doyle, *Lines Across the Land*, which includes case studies of rural electric cooperatives in fourteen states (Alabama, Georgia, Illinois, Indiana, Kentucky, Michigan, Minnesota, Montana, Nebraska, North Dakota, South Dakota, Virginia, Wisconsin, and Wyoming).

66 **Letter from IREA urging support for Patrick Michaels:** "Helping a Global Warming Skeptic," *Inside Higher Education*, July 31, 2006.

67 **Congressman Jim Cooper of Tennessee wrote:** Jim Cooper, "Electric Co-operatives: From New Deal to Bad Deal?" *Harvard Journal on Legislation*, 2008.

68 **Early organizing by the Sierra Club on the Sunflower project, project hearings, and development of the coalition opposing the project:** Scott Martelle, "Killing King Coal," *Sierra Magazine*, March/April 2009.

EIGHT

Direct Action

73 **In Montana, Oklahoma, Kentucky, and Michigan, judges and regulators handed out rejection slips to coal plants:** Roundup Power Project (Montana), Red Rock Generating Station (Oklahoma), Thoroughbred Generating Station (Kentucky), Escanaba plant (Michigan). For details, see Appendix B, "Coal Plants Canceled, Abandoned, or Put on Hold."

73 **In North Dakota, Arizona, Washington, and New York, companies withdrew projects on their own initiative:** South Heart Power Project (North Dakota), Bowie IGCC Power Station (Arizona), Avista plant (Washington), Russell Station II (New York). For details, see Appendix B, "Coal Plants Canceled, Abandoned, or Put on Hold."

73 **Citigroup downgraded the stocks of mining companies:** Jim Jelter, "Coal Stocks Tumble on Citigroup Downgrade," *MarketWatch*, July 18, 2007.

73 **In August U.S. Senate Majority Leader Harry Reid became the highest-ranking federal official to speak out against the building of coal-fired power plants:** Sean Whaley, "Clean Energy Summit: Reid Opposes Coal-Fired Power

Plants, Senate Majority Leader Says Sun Is an Untapped Resource for the Silver State," *Las Vegas Review-Journal,* August 19, 2007.

74 **Ollie Combs:** Chad Montrie, *To Save the Land and People: A History of Opposition to Surface Coal Mining in Appalachia* (University of North Carolina Press, 2002). The account of Ollie Combs and others in Knott County is told in Chapter 5, "We Will Stop the Bulldozers: Opposition to Surface Coal Mining in Kentucky, 1967–1972."

74 **Knott County strip mine occupation:** Stephen L. Fisher, editor, *Fighting Back in Appalachia* (Temple University Press, 1993).

75 **Knott County and Perry County sabotage; Mountaintop Gun Club:** Ibid.

75 **Minnesota power line uprising:** Paul Wellstone and Barry M. Casper, *Powerline: The First Battle in America's Energy War* (University of Minnesota Press, 2003).

75 **Rocky Top affinity group action at Zeb Mountain:** john johnson, "Don't Chop Rocky Top: Katúah EF! Confronts Mountaintop Removal in Tennessee," *Earth First! Journal,* Samhain/Yule, 2003.

76 **Agnone paper:** Jon Agnone, "Amplifying Public Opinion: The Policy Impact of the U.S. Environmental Movement," *Social Forces,* June 2007.

78 **Radical cheerleading:** For more on radical cheerleading and other creative activist tactics, see Mike Hudema, *An Action a Day Keeps Global Capitalism Away* (Between the Lines, 2004).

79 **"I hope the impatience..."** Floyd Lilley, "Super Facilitator Pronk Makes All Share Pain," *WorldNetDaily,* November 24, 2000, sovereignty.net/p/clim/haguc1100/updates.html.

80 **Inside/outside tactics:** For an extended discussion of inside/outside tactics, see Joshua Kahn-Russell, "Climate Justice and Coal's Funeral Procession," *Z Magazine,* May 2009.

82 **"High school and college students are red meat for banks."** Mike Brune, *Coming Clean: Breaking America's Addiction to Oil and Coal* (Sierra Club Books, 2008), 82.

83 **In Knoxville, Tennessee, police had used choke holds and pain compliance:** "First stockholders meeting of National Coal Corporation Disrupted," Mountain Justice website, April 29, 2008.

83 **Company workers had threatened protesters and attempted to ram them with a car:** "Eleven Arrested Protesting ET Coal Mine," WBIR website, August 15, 2005; "Tennessee Coal Road Blocked to Protest Mountaintop Removal Mining," Mines & Communities website, August 15, 2005.

83 **In North Carolina, protesters at Dominion's Cliffside Plant were tasered and placed in pain compliance holds.** "Eight Climate Protesters Arrested at U.S. Coal Plant," Reuters, April 1, 2008; "Eight Arrested as North Carolina

Residents Shut Down Construction at Cliffside Coal Plant," Fossil Fools Day blog, April 1, 2008.

83 **In Ohio, police pepper-sprayed protesters conducting a sit-in at the headquarters of American Municipal Power:** "Women Climb Flagpole In Power Plant Protest," NBC 4 Columbus, July 7, 2008; "Police arrest protesters at Ohio power company," WDTN 2 Dayton, July 8, 2008.

83 **In West Virginia, mine workers threatened and assaulted anti-coal activists:** Dave Cooper, "West Virginia Coal Thugs Disrupt July 4th Picnic (Video)," *Huffington Post*, July 7, 2009.

83 **Houses of activists were been shot at, vandalized, and even fire-bombed:** Kurt Pitzer, "Last Man on the Mountain," *People Magazine*, October 20, 2008; Jeff Goodell, *Big Coal: The Dirty Secret Behind America's Energy Future* (Houghton-Mifflin 2006), Chapter 2, "Coal Colonies."

83 **Charges against Hannah Morgan and Kate Rooth:** James Hansen, "Obstruction of Justice: Virginian Coal Protesters Receive B-Minus Plea Bargain for Kingsnorth-Like Activism," *Grist*, October 27, 2008.

84 **"If this case had gone to trial..."** James Hansen, "Obstruction of Justice," *Grist*, October 27, 2008.

NINE

The Education of Warren Buffett

This chapter is based on Ted Nace, "The Education of Warren Buffett: Why Did the Guru Cancel Six Coal Plants?" *Grist*, April 15, 2008.

85 **"Nothing is illegal if 100 businessmen decide to do it."** Ironically, the quote was originally part of an assertion by Andrew Young that big business might play a progressive role in ending the apartheid system in South Africa, as recounted in Joseph Lelyveld, "New Voice of U.S. At United Nations," *Ocala Star-Banner*, February 14, 1977, 7D:

> Surprisingly, in Young's vision, the catalyst that brings about change turns out to be that troublesome and maligned behemoth, the American multinational corporation. Thinking aloud one afternoon, Young wondered what would happen if the U.S. Government urged American corporations in South Africa to act on the assumption that they had five years to turn their management over to blacks. Hundreds of blacks would then have to be brought to the United States for management training, he said, imagining how the thin edge of this wedge might fit into the locked door of apartheid. Such a scheme might prove to be illegal under South African law, I pointed out. "I know," Young answered, smiling, "but nothing is illegal if 100 businessmen decide to do it, and that's true anywhere in the world."
>
> Young regularly summons "100 businessmen" into his conversation in

this way, whenever he wants to talk about moral suasion as a practical exercise. They are a kind of collective stock character but not entirely of his own imagining. In 1963, in Birmingham, Ala., the Rev. Andrew Young negotiated with 100 or so white businessmen to bring about the dismantling of the segregation laws. Atlanta was an even better example of reformist oligarchy. If Atlanta became a city "too busy to hate," Young seems to be saying, then why not Johannesburg?

"My notion is," he said, "that if revolution is the transfer of goods, services, and opportunities, then capitalism has produced a lot more in the way of revolution than Communism." Whether we like it or not, he argued, the multinationals involve the United States in the affairs of other countries. That being so, he continued, "Why not incorporate a sense of political direction with the profit motive?"

86   **Table 1:** "Key private sector decision makers on coal," CoalSwarm wiki.

87   **"CEOs of fossil energy companies..."** James Hansen: "Try Fossil Fuel CEOs for 'High Crimes Against Humanity'," *Environmental Leader,* June 24, 2008.

87   **Hansen letter to James E. Rogers:** Hansen's commentaries, articles, letters to prominent businessmen and government officials, and papers can be found on his Columbia University website at www.columbia.edu/~jeh1/.

89   **MidAmerican Energy's operations:** "MidAmerican Energy," CoalSwarm wiki.

89   **At least four others would be built in the Rocky Mountain region:** In the fall of 2006, PacifiCorp developed 12 scenarios, which are shown in Table 7.37 of the company's *2007 Integrated Resource Plan* (May 30, 2007 release). Of these scenarios, 8 showed the addition of 6 new coal plants in the 2012–2018 period, while 4 showed the addition of 7 new coal plants.

89   **Plans to build new coal plants were off the table:** Dustin Bleizeffer, "Utility snuffs coal projects," *Casper Star-Tribune,* December 11, 2007.

91   **Four coal plants ... were now omitted:** In the May 30, 2007 release of the *2007 Integrated Resource Plan,* the scenario omitting four coal plants is referred to as "Group 2." The company explained: "The feedback received on the [Group 1] resource proposal, as well as recent external events and an assessment of state resource policy directions, prompted the company to investigate portfolio alternatives that recognize existing and expected state resource acquisition constraints." PacifiCorp, *2007 Integrated Resource Plan,* 153.

91   **"Oregon PUC rejection of the 2012 RFP...":** PacifiCorp, *2007 Integrated Resource Plan,* 153.

91   **"[W]hat we are really talking about with global warming is dislocation..."** David Neubert, "Berkshire Hathaway on Global Warming," ThePanelist.com, May 5, 2007.

92   **"Well, fortunately climate change, although it's a huge challenge..."** Transcript of remarks by Bill Gates, Chairman, Microsoft Corporation; Craig

Mundie, Chief Research and Strategy Officer, Microsoft Corporation; Professor Muhammad Yunus, Founder—Grameen Bank, 2006 Nobel Peace Prize Laureate; Microsoft Government Leaders Forum—Asia 2007 at Beijing, People's Republic of China, April 19, 2007; transcript posted on Microsoft Web site, microsoft.com/Presspass/exec/billg/speeches/2007/04-19GLF_Asia_Keynote.mspx.

92 **"...perhaps a foot and a half."** This estimate of sea level rise is consistent with projections by the Intergovernmental Panel on Climate Change (IPCC): 7 to 23 inches by the end of the century. But the IPCC projection is generally regarded as conservative, since it does not include the impact on sea levels of the rapid ice sheet collapses that have accompanied previous meltings. For example, at the end of the last ice age about 14,000 years ago, sea levels rose an average of three feet every two decades for 400 years for a total of 60 feet. In his book *Censoring Science,* Mark Bowen notes: "The first three feet would be sufficient, incidentally, for the entire Mississippi River Delta, including New Orleans, to vanish into the Gulf of Mexico; for tens of millions of people in Bangladesh, one of the most densely populated regions on Earth, to be forced to migrate; and for rice-growing river deltas throughout Asia, a major source of food for the human species, to be inundated." Mark Bowen, *Censoring Science* (Dutton, 2007), 109.

92 **"Within the last few months..."** Dustin Bleizeffer, "Utility snuffs coal projects," *Casper Star-Tribune,* December 11, 2007.

94 **Text of Alexander Lofft petition:** thepetitionsite.com/takeaction/6617214 06#body.

95 **"I'll tell you why I like the cigarette business..."** Jennell Wallace, "Warren Buffett cools on his attraction to tobacco business," *Bloomburg News,* April 25, 1994.

95 **"fraught with questions..."** Ibid.

TEN

Progress Report: 59 Coal Plants Down

97 **Opposition by mayors:** Additional details: On December 19, 2007, Charlottesville passed the Charlottesville Clean Energy Resolution putting the city on record in support a moratorium. On October 13, 2007, Pocatello, Idaho, mayor Roger Chase told other mayors from across the state attending an Association of Idaho Cities legislative committee that he favored a moratorium on new coal plants in the state. On June 1, 2007, Park City, Utah, mayor Dana Wilson wrote a letter to Warren Buffett expressing the city's opposition to three coal plants proposed by Rocky Mountain Power. In November 2007, Salt Lake City mayor Rocky Anderson expressed his support for a coal moratorium at a rally

organized by the Step It Up! campaign. Source: "Coal moratorium," CoalSwarm wiki.

98 **"Coal is a double-edged sword."** Steven Mufson, "Coal Rush Reverses, Power Firms Follow," *Washington Post,* September 4, 2007

98 **Protests and referendum in Wiscasset, Maine:** "Lobstermen Protest Proposed Power Plant," WCSH Portland website, October 11, 2007. For further details see "Maine and coal," CoalSwarm wiki.

98 **Kalama Plant:** Erik Robinson, "State Rejects Proposal for Coal Plant in Kalama," *Clark County Columbian,* November 28, 2007; Erik Robinson, "Power Plant Plant to Drop New Coal Technology," *Vancouver Columbian,* December 23, 2007; Max Zygmont, "Energy Northwest Walks Away From Pacific Mountain Energy Center," Climate Change and Carbon Management blog, September 23, 2008. See also "Pacific Mountain Energy Center" and "Washington and coal," CoalSwarm wiki.

99 **Protests against Fisk and Crawford:** Extensive background information on organizing efforts against these two Chicago plants may be found on the websites of LVEJO (Little Village Environmental Justice Organization) and PERRO (Pilsen Environmental Rights and Reform Organization PERRO). Also see "Fisk Generating Station," "Crawford Generating Station," and "Illinois and coal," CoalSwarm wiki.

99 **Harvard School of Public Health:** Jonathan I. Levy et al., "Using CALPUFF to evaluate the impacts of power plant emissions in Illinois: model sensitivity and implications," *Atmospheric Environment,* 2002.

100 **Step It Up!** Manny Fernandez, "For the Environment, Rallies Great and Small (and Unusual Attire)," *New York Times, April 15, 2007.* For details on November 2007 Step It Up! actions across the United States, see events.stepitup2007.org/november/reports.

101 **ten cancellations of "clean coal" projects:** Steve Raabe, "'Clean coal' plant setbacks mount in U.S.," *Denver Post,* November 1, 2007.

102 **seventeen U.S. coal plants had been canceled:** "The Growing Trend Against Coal-Fired Power Plants (USA)," Palang Thai website, December 1, 2007.

105 **our list of projects ... had grown to fifty-nine:** Matt Leonard, "2007: A Rough Year for Coal," *It's Getting Hot In Here,* January 17, 2008. For the complete list, see "Coal plants cancelled in 2007," CoalSwarm wiki. Also see Appendix B, "Coal Plants Canceled, Abandoned, or Put on Hold."

105 **"This is part of a concerted effort to grossly exaggerate opposition..."** Steve James, "Coal's time is up, say environmentalists," *World Environment News,* February 15, 2008.

106 **"Reports of this project's demise are greatly exaggerated."** David Bertola, "Is Huntley project on hold—or simply delayed?" *Buffalo Business First,* January 18, 2008.

106 **"What began as a few local ripples…"** Lester Brown, "U.S. Moving Toward Ban on New Coal-Fired Power Plants," Earth Policy Institute, February 14, 2008.

107 **In a September 2007 national poll … only 3 percent chose coal:** Opinion Research Corporation, "A Post Fossil-Fuel America," Executive Summary, National Opinion Survey Produced for Citizens Lead for Energy Action Now (CLEAN), A Project of the Civil Society Institute, October 18, 2007.

107 **Financial problems with nuclear power:** Jay M. Gould, "The future of nuclear power," *Monthly Review*, February 1984.

ELEVEN
Unicorns, Leprechauns, Clean Coal

112 **75 percent of the public supporting a five-year moratorium:** Opinion Research Corporation, "A Post Fossil-Fuel America," Executive Summary, National Opinion Survey Produced for Citizens Lead for Energy Action Now (CLEAN), A Project of the Civil Society Institute, October 18, 2007.

112 **Bob Henrie:** "Bob Henrie," CoalSwarm wiki on SourceWatch; Mike Gorrell, "Keeping coal in the Spotlight: High-Powered P.R. Firm Goes to Work Polishing the Image," *Salt Lake Tribune*, March 1, 2008; Joe Geiser and Cecelia Moriarity, "Company Greed Killed Coal Miners in Utah: 20 Years Since the Wilberg Mine Disaster; How Emery Mining Corp. Tried to Hide the Facts," *The Militant*, December 28, 2004.

113 **"Presidential Race Runs through the Heart of Coal County…"** Reuters, October 10, 2008.

113 **Memos leaked to the press:** Kevin Grandia, "Coal lobby PR firm memo boasts about manipulating Democrats and Republicans," *DeSmogBlog*, January 16, 2009; "American Coalition for Clean Coal Electricity" and "Clean Coal Marketing Campaign," CoalSwarm wiki.

114 **"The 'clean coal' PR people are running a scam…"** David Roberts, "The essential 'clean coal' scam: Politico lets shill get away with the basic dodge at the center of the 'clean coal' campaign," *Grist*, December 23, 2008.

114 **"the earliest possible deployment … is not expected before 2030."** Emily Rochon, *False Hope: Why Carbon Capture and Storage Won't Save the Climate*, Greenpeace, 2008.

115 **Highwood pollution figures:** The figures for the Highwood Power Project are from "Cleanest in the Country? Highwood and Pollution," Montana Environmental Information Center, meic.org/energy/power_plants/highwood/highwood-facts. Notes on the public health impacts of pollutants are from "Environmental

Impacts Of Coal: Air Pollution," Union of Concerned Scientists, ucsusa.org/
clean_energy/coalvswind/co2c.html.

119 **ACCCE made a $35 million commitment:** Steven Mufson, "Coal Industry
Plugs Into the Campaign," *Washington Post,* January 18, 2008.

T W E L V E
War Against the Mountains

124 **union organizers … arrested, beaten, and even killed.** Robert Justin Goldstein,
Political Repression in Modern America: From 1870 to 1976 (University of Illinois
Press, 1978, 2001), 189.

125 **In 1921 … thirteen thousand West Virginia coal miners:** Robert Shogan, *The
Battle of Blair Mountain: The Story of America's Largest Labor Uprising* (West-
view Press, 2004); Jeff Biggers, *The United States of Appalachia* (Shoemaker &
Hoard 2006); Lon Savage, *Thunder in the Mountains: The West Virginia Mine
War, 1920–21* (University of Pittsburgh Press 1990).

126 **Formation of Save Our Cumberland Mountains (1972):** Chad Montrie, *To
Save the Land and People: A History of Opposition to Surface Coal Mining in
Appalachia* (The University of North Carolina Press, 2003), 186.

126 **Weakness of the Surface Mining Control and Reclamation Act; effect on the
anti-mining movement:** Chad Montrie, *To Save the Land and People,* Chapter 9:
"Against the Little Man Like Me: Legalized Destruction in the SMCRA Era."

126 **The first such operation began in 1970:** Ken Ward, Jr., "Flattened: Most
mountaintop mines left as pasture land in state," *The Charleston Gazette,* August
9, 1998.

126 **No town was ever actually built:** Ken Ward Jr., "Flattened: Most mountaintop
mines left as pasture land in state," *Sunday Gazette-Mail,* August 9, 1998.

127 **James "Buck" Harless and the 2000 U.S. presidential election:** "James Harless," CoalSwarm wiki; Thomas B. Edsall, Sarah Cohen and James V. Grimaldi,
"Pioneers fill war chests, then capitalize," *Washington Post,* May 16, 2004.

128 **"payback" from the new administration:** The full quote is: "You did everything
you could to elect a Republican president. You are already seeing in his actions
the payback, if you will, his gratitude for what you did." Craig Aaron, "Bought
and Paid For," *In These Times,* March 26, 2004.

128 **Reclassification of "waste" to "fill":** Joby Warrick, "Appalachia Is Paying Price
for White House Rule Change," *Washington Post,* August 17, 2004.

132 **Downstream Strategies wind study:** Evan Hansen et al., *The Long-Term
Economic Benefits of Wind Versus Mountaintop Removal Coal on Coal River
Mountain, West Virginia* (Downstream Strategies, 2008).

132 **Clean Water Protection Act:** The bill would add the following paragraph to Section 502 of the Federal Water Pollution Control Act:

> Fill Material.—The term "fill material" means any pollutant which replaces portions of the waters of the United States with dry land or which changes the bottom elevation of a water body for any purpose. The term does not include any pollutant discharged into the water primarily to dispose of waste.

133 **"As I write this letter, I brace myself..."** This is an abridged version of Bo Webb's letter. The complete text can be found at Dave Cooper, "Letter From a Mountaineer to President Obama," *Huffington Post,* February 23, 2009.

134 **By some estimates, only ten to twenty years of economically minable coal remained in Appalachia.** Leslie F. Ruppert, *Resource Assessment of Selected Coal Beds and Zones in the Northern and Central Appalachian Basin Coal Regions* (USGS Professional Paper 1625-C, 2000): "Sufficient high-quality, thick, bituminous resources remain in [Appalachian Basin] coal beds and coal zones to last for the next one to two decades at current production."

THIRTEEN

The Grandmother Rebellion

136 **Background on the history of Black Mesa:** Judish Nies, "Black Mesa Syndrome: Indian Lands, Black Gold," *Orion Magazine,* Summer 1998.

137 **"Somewhere far away from us, people have no understanding..."** Sean Patrick Riley, "Gathering Clouds," *Los Angeles Times,* June 6, 2004.

137 **During his four years working, he had to fly five children:** Peter Montague, "Resistance at Desert Rock, *Rachel's Democracy and Health News* #889, January 11, 2007.

137 **More than five times as likely to be seen ... for respiratory complaints:** "Navajo Coal and Air Quality in Shiprock, New Mexico," USGS Fact Sheet 2006-3094, July 2006.

137 **"as bright as daffodils"** Peter Montague, "Resistance at Desert Rock, *Rachel's Democracy and Health News* #889, January 11, 2007.

138 **The American Lung Assocation estimates 16,000 people:** Jeff Conant, "Speaking Diné to Dirty Power: Navajo Challenge New Coal-Fired Plant," *CorpWatch,* April 3, 2007.

138 **San Juan and Four Corners emissions:** Ibid.

138 **150 million tons of coal combustion waste:** Ibid.

138 **In 1974 John Boyden and his coal industry allies pushed legislation through Congress:** Judith Nies, "Black Mesa Syndrome: Indian Lands, Black Gold," *Orion Magazine,* Summer 1998.

139 **"...one of the worst cases of involuntary community resettlement that I have studied."** Letter from Thayer Scudder, Professor of Anthropology, California Institute of Technology, to Abdelfattah Amor, Special Rapporteur of the United Nations Commission on Human Rights, January 30, 1998.

139 **"I feel that in relocating these elderly people, we are as bad as the Nazis..."** *Women Transform The Mainstream: 18 Case Studies of Women Activists Challenging Industry, Demanding Clean Water and Calling for Gender Equality in Sustainable Development,* Sixth Session of the United Nations Commission on Sustainable Development (CSD-6), 1998.

139 **Enticed by the promise of a $50 million annual payout to the Navajo Nation, the Tribal Council voted 66-7 in favor of inviting Sithe to build the plant:** "Navajo Speaker Morgan Reaffirms Commitment to Desert Rock," Desert Rock Energy Company press release, September 16, 2007.

140 **Ecos Consulting study:** *Renewable Energy and Economic Alternatives for the Navajo Nation* (Ecos Consulting, 2008).

141 **Desert Rock blockade:** Kathy Helms, "Protesters Blockade Desert Rock Site," *Gallup Independent,* December 13, 2006; Kathy Helms, "Resisters Move: No Arrests Made at Desert Rock's Future Location," *Gallup Independent,* December 22, 2006; Kathy Helms, "Spiritual Gathering to Heal the Earth Begins Thursday," *Gallup Independent,* November 7, 2007.

141 **"They had a small, white tent...."** Virginia L. Clark, "Resistance Leads to Human Blockade for Relatives of Taos Poets," *Call of the Wild,* January 1, 2007.

141 **Scores of local residents expressed vehement opposition:** Jason Begay, "Desert Rock Critics Flood Final Hearing," *Navajo Times,* September 26, 2007.

141 **Diné CARE sued the federal Office of Surface Mining:** Joe Hanel, "Judge Sets Timetable for Desert Rock Case," *Durango Herald,* December 8, 2007.

141 **Governor Richardson opposes plant:** Sue Major Holmes, "Richardson Speaks Out Against Coal Power Plant," Associated Press, July 27, 2007.

141 **Mountain Ute Tribal Council resolution:** Chuck Slothower, "Mountain Utes Oppose Desert Rock," *Durango Herald,* August 25, 2007.

141 **EPA expresses concerns:** "Desert Rock Energy Project: Power Plant Document Concerns EPA," *Santa Fe New Mexican,* September 16, 2007.

142 **Diné CARE presented Sithe with a report:** *Renewable Energy and Economic Alternatives for the Navajo Nation* (Ecos Consulting, 2008).

142 **Just Transition Coalition:** "Just Transition Coalition" on CoalSwarm wiki at SourceWatch.org; Christopher McLeod, "Seeking a Just Transition," *Earth Island Journal,* Summer 2006.

143 **Steve Schwarzman 99 percent pay cut and Blackstone $1.33 billion loss:** "Blackstone CEO Takes 99 Percent Pay Cut," Reuters, March 3, 2009.

143 **Price escalation from $1.5 billion to $4 billion:** Will Sands, "Captured at Desert Rock: Power Plant Appeals for Federal Funds," *Durango Telegraph*, August 13, 2009.

144 **Arizona Public Service stated its intention to move away from coal:** Ryan Randazzo, "2nd APS Solar Plant May Surpass Target Set for Green Energy," *The Arizona Republic*, May 22, 2009.

144 **$110,000 in legal fees:** "Discussion Points on the Desert Rock Power Plant," Diné Citizens Against Ruining Our Environment, February 26, 2009.

144 **Joe Shirley Jr. continued to win key Tribal Council votes:** "Navajo President Joe Shirley, Jr., Pleased by Council Vote to Approve Desert Rock Energy Project Right-of-Way Permit," Navajo Nation press release, February 28, 2009.

144 **"I worked on plenty of power plants..."** *Renewable Energy and Economic Alternatives for the Navajo Nation* (Ecos Consulting, 2008).

FOURTEEN
Cowboys Against Coal

146 **All ten of the largest coal mines:** "Major U.S. Coal Mines," Energy Information Administration, eia.doe.gov/cneaf/coal/page/acr/table9.html.

146 **More coal than the entire state of West Virginia:** In 2007 West Virginia produced 153 million short tons of coal. North Antelope Rochelle produced 92 million short tons, and Black Thunder produced 86 million short tons. "Production and Number of Mines by State and Mine Type," Energy Information Administration, eia.doe.gov/cneaf/coal/page/acr/table1.html; "Major U.S. Coal Mines," Energy Information Administration, eia.doe.gov/cneaf/coal/page/acr/table9.html.

146 **According to some estimates they nearly equal the reserves of China:** Although China officially placed its coal reserves at 115 billion metric tons in 1992, California Institute of Technology professor David Rutledge estimates that the country's remaining reserves had declined to about 71 billion metric tons as of 2006. Montana's remaining reserves were 67.9 billion metric tons (74.9 billion short tons) in 2006, according to the Energy Information Administration. Source: David Rutledge, *Hubbert's Peak, The Coal Question, and Climate Change*, Watson Lecture at Caltech, October 2007, rutledge.caltech.edu/. For further details, see "Coal reserves," CoalSwarm wiki.

147 **North Central Power Study:** The study was released by the U.S. Bureau of Reclamation. Its conclusions are summarized in K. Ross Toole, *The Rape of the Great Plains: Northwest America, Cattle and Coal* (Atlantic Monthly Press and Little, Brown, 1976), 19–20.

147   **Founding of Northern Plains Resource Council; description of Boyd Charter:** K. Ross Toole, *The Rape of the Great Plains: Northwest America, Cattle and Coal* (Atlantic Monthly Press and Little, Brown, 1976), 122, 222–223.

149   **Opposition to Highwood plant in Helena and Missoula:** See Chapter Ten, "Progress Report: 59 Coal Plants Down."

149   **RUS loan moratorium:** Letter from James M. Andrew, Administrator, United States Department of Agriculture Utilities Programs to Tim Gregori, Southern Montana Electric Generation and Transmission Cooperative, February 19, 2008.

149   **BER ordered more research on particulates smaller than 2.5 microns:** "State Orders More Study of Emissions," *Great Falls Tribune*, April 22, 2008.

150   **Ron Sega:** "Rentech Announces Appointment of Former U.S. Air Force Under Secretary Ron Sega to Its Board of Directors," *Business Wire*, December 18, 2007.

150   **Air Force synfuels plans:** Dave Montgomery, "Air Force pushing for liquefied coal to power its fleet," *Lawrence Journal-World*, March 30, 2008; "Air Force Wants Coal for Fuel, But Will Idea Fly?" Associated Press, March 31, 2008.

151   **Air Force synfuels lobbying:** "Rentech Hires Lobbying Firm," Associated Press, January 2, 2008.

151   **Malmstrom project:** "Malmstrom Air Force Base Coal-to-Liquids" and "U.S. Air Force and Coal," CoalSwarm wiki.

151   **Waxman/Davis letter to Gates:** oversight.house.gov/documents/20080130112607.pdf.

152   **Air Force cancels Malmstrom:** Peter Johnson, "Officials Scrap Plans for Plant at Malmstrom," *Great Falls Tribune*, January 30, 2009.

152   **Background on North Dakota:** "North Dakota and coal," CoalSwarm wiki.

153   **Five North Dakota plants among the 50 worst emitters of carbon dioxide and mercury:** Measured in pounds of carbon dioxide or mercury per megawatt-hour. *Dirty Kilowatts: America's Most Polluting Plants* (Environmental Integrity Project, July 2007).

153   **"The Bush gang finally gave up and we did not."** Terrence Kardong, "A Win for the Mouse," *Dakota Counsel*, December 10, 2008.

154   **Big Stone II chronology:** "Big Stone II," CoalSwarm wiki.

154   **Eight Minnesota legislators:** Ron Way, "Big Stone II project faces increasing scrutiny," *Minnesota Post*, April 29, 2008.

154   **David Schlissel testimony:** *Direct Testimony of David A. Schlissel and Anna Sommer, Synapse Energy Economics, Inc.,* (South Dakota Public Utilities Commission Case No EL05-022, May 19, 2006).

155 **Leslie Glustrom's findings on limits to coal supplies:** Leslie Glustrom, *Coal Supply Constraints: Cheap & Abundant, Or Is It?* (Colorado Clean Energy Action, 2009).

156 **USGS study of the Gillette field:** James Luppens et al., *Assessment of Coal Geology, Resources, and Reserves in the Gillette Coalfield, Powder River Basin, Wyoming* (US Geological Survey Open-File Report 2008-1202, 2008).

156 **Two Big Stone II cosponsors withdrew:** Carissa Wyant, "Utilities Withdraw from Power Plant Project," *Minneapolis-St. Paul Business Journal,* September 18, 2007.

156 **Utilities underestimated construction costs:** "Big Stone II Power Plant Partners Study Independent Report," *West Central Tribune Online,* October 24, 2008.

156 **EPA filed objections to Big Stone II's air permit:** "Big Stone II Sent Back to the Drawing Board," Clean Water Action website, January 23, 2009.

FIFTEEN

Sierra

158 **Bruce Nilles:** Samara Kalk Derby, "A Season to Fight Coal: Madison Lawyer Piles Victories," *Madison Capital Times,* December 14, 2007.

159 **Development of Sierra Club National Coal Campaign:** Scott Martelle, "Killing King Coal," *Sierra Magazine,* March/April 2009; Matthew Brown, "Environmentalists Lead Charge Against Coal," Associated Press, January 15, 2008; *Sierra Club Foundation Annual Report 2008.*

160 **Erik Shuster's list of 151 proposed new coal plants:** *Tracking New Coal-Fired Power Plants* (National Energy Technology Laboratory, May 31, 2007) is archived at http://cmNOW.org. The most release of the report is posted at netl.doe.gov/coal/refshelf/ncp.pdf.

160 **Of the twenty-four thousand people estimated to die prematurely in the United States due to fine particles from power plants, a third are in the six industrial heartland states:** *Dirty Air, Dirty Power,* Clean Air Task Force, June 2004. The study estimates 7546 annual mortalities from power plant pollution in the following states: Pennsylvania 1825, Ohio 1743, Illinois 1356, Michigan 981, Indiana 887, Missouri 754.

160 **Illinois:** Peter Downs, "Addicted to Coal: The Battles Being Waged Here Will Shape the Nation's Energy Debate," *Illinois Times,* April 17, 2008; "Illinois and coal," CoalSwarm wiki.

160 **EnviroPower:** "Franklin County Power Plant," CoalSwarm wiki.

160 **"The Sierra Club's latest salvo..."** Peter Downs, "Addicted to Coal: The Battles Being Waged Here Will Shape the Nation's Energy Debate," *Illinois Times*, April 17, 2008.

161 **Taylorville Energy Center:** "Taylorville Energy Center," CoalSwarm wiki.

161 **Madison plants:** Samara Kalk Derby, "A Season to Fight Coal: Madison Lawyer Piles Victories," *Madison Capital Times*, December 14, 2007; "Stopping the Coal Rush," Sierra Club new coal plant proposal tracking list; "Doyle: No Coal at UW by 2012," *Badger Herald*, February 8, 2009; "Wisconsin and coal," CoalSwarm wiki; "Charter Street Heating Plant," CoalSwarm wiki.

162 **Texas, Missouri, Florida...** For a frequently updated status list of Sierra Club opposition to new coal plants across the country, see "Stopping the Coal Rush," sierraclub.org/environmentallaw/coal/plantlist.asp.

162 **Sierra's Dynegy campaign** Ted Nace, "Out of the Frying Pan: Dynegy Targeted by Sierra Club in New Anti-Coal Campaign," *Grist*, April 8, 2008; "Dynegy / LS Power," CoalSwarm wiki.

162 **In May, a hundred Sierra activists showed up at Dynegy's annual meeting and delivered 10,000 letters and emails:** Scott Martelle, "King Coal in Court," *Sierra Magazine*, May/June 2009.

164 **"Very little new power plant development is going on..."** Dynegy Inc. Q3 2008 Earnings Call Transcript, *Seeking Alpha*, November 6, 2008.

165 **Deseret petition:** Scott Martelle, "King Coal in Court," *Sierra Magazine*, May/June 2009.

166 **Seattle hearing:** Heather M., "Report From the Seattle EPA Global Warming Hearing," *Climate Crossroads*, May 22, 2009.

SIXTEEN

Taking It to the Streets

168 **Chris Copeland's account of the Harriman spill:** Chloe White, "TVA Ash Pond Breach: Resident Says Area Has 'Changed Forever'," *Knoxville News Sentinel*, December 23, 2008.

168 **It was one hundred times the reported size of the Exxon Valdez disaster:** "Tennessee sludge spill estimate grows to 1 billion gallons," CNN, December 26, 2008. The "one hundred times" figure is based on the eleven million gallon figure commonly cited for the size of the Exxon Valdez spill. In her account of the Exxon Valdez spill *Not One Drop* (Chelsea Green, 2008, 193–194), marine biologist Dr. Riki Ott points out that Exxon actually estimated the volume of the spill at between eleven and thirty-eight million gallons, and the state of Alaska estimated the spill at thirty to thirty-five million gallons.

169 **Twitter was abuzz:** Sandra Diaz, "New Media Keeping Coal Ash Spill From Drowning in the Muck," *Huffington Post,* January 8, 2009. According to Diaz, other crucial work in the early days to publicize the spill was done by the group United Mountain Defense, by Dave Cooper of the Mountaintop Removal Road Show, and by the Sierra Club.

169 **A Google search for the phrase "Tennessee spill":** Ted Nace, "Blowback: Did the Coal Industry Create Its Own PR Nightmare?" *Grist,* December 31, 2008.

169 **"My worst nightmare":** Steven Chu presentation at U.C. Berkeley, April 23, 2007. The quote, in context, is as follows: "Let me go to the supply side of the energy problem. Now, we have lots of fossil fuel. That's really both good and bad news. We won't run out of energy, but there's enough carbon in the ground to really cook us. Coal is my worst nightmare. Carbon emission in the next thirty years is predicted in the current forecast to—we'll be adding three times the amount of carbon dioxide in the previous history of all humanity if we continue on our present course." Page Van Der Linden, "Nightmare on Coal Street: The Video," *DeSmogBlog,* December 19, 2008.

171 **Capitol Power Plant:** Ted Nace, "Mean, Old, and Dirty: Climate Youth Activists Target the Capitol Power Plant," *Grist,* December 24, 2008; "Capitol Power Plant," CoalSwarm wiki; Jim Spellman and Andrea Koppel, "Effort to 'green' U.S. Capitol complicated by coal," CNN.com, May 11, 2007; "The Capitol Power Plant," *Hill Rag,* January 2006; Lyndsey Layton, "Capitol Hill's polluting power plant resists green tide," *Washington Post,* April 26, 2007; Lyndsey Layton, "Reliance on Coal Sullies 'Green the Capitol' Effort," *Washington Post,* April 21, 2007.

173 **"We aim to create an action framework..."** Ted Nace, "Mean, Old, and Dirty: Climate Youth Activists Target the Capitol Power Plant," *Grist,* December 24, 2008.

174 **"We can determine the fate of our generation..."** Ibid.

175 **"There are moments in a nation's—and a planet's—history..."** Mike Brune, "Wendell Berry and Bill McKibben Call for Mass Civil Disobedience Against Coal," *Huffington Post,* December 12, 2008.

176 **Ayers reported that the switch had been accomplished:** Jordy Yager, "Capitol Power Plant to Stop Use of Coal," *The Hill,* May 1, 2009.

177 **Following the Capitol Climate Action, the number of direct action protests against coal immediately increased both in frequency and size:** See Appendix A: "Protests Against Coal."

180 **Assuming an average lifespan of fifty years, those 101 plants would have emitted 20 billion tons of carbon dioxide:** "100 Coal Plants Prevented or Abandoned: Movement Sparks Shift to Cleaner Energy and Over 400 Million Fewer Tons of CO_2," Sierra Club press release, July 9, 2009. The Sierra estimate of 400 million tons of CO_2 per year is derived from the CARMA database of

power plants and proposed power plants. Details on the cancelled plants may be found on Sierra's "Stopping the Coal Rush" online tracking list of power plants at sierraclub.org/environmentallaw/coal/plantlist.asp.

180 **Virgin Earth Challenge:** Ted Nace, "Sir Richard Branson: Hand Over the $25 Million! Why the No New Coal Plants Movement Should Be Awarded the Virgin Earth Challenge Prize," *Grist,* January 2, 2009.

181 **"In a few years, the backlash against coal power..."** Juliette Jowit, "Coal Plans Go Up in Smoke," *Manchester Guardian,* September 3, 2008.

APPENDIX A

Protests Against Coal

For further details on particular protests as well as information on protests that have taken place subsequent to the publication of this book, see "Nonviolent direct actions against coal," CoalSwarm wiki, sourcewatch.org/index.php?title=Nonviolent_direct_actions_against_coal.

183 **Blockade at Zeb Mountain:** "Bannerhang and Blockade," Mountain Justice website, accessed April 2009; john johnson, "Don't Chop Rocky Top: Katúah EF! Confronts Mountaintop Removal in Tennessee," *Earth First! Journal,* Samhain/ Yule, 2003.

183 **Chesapeake Climate Action Network blockade of Dickerson Power Plant:** "Demonstrators Decry Mirant Corporation for Ignoring Public Health and Global Warming," Chesapeake Climate Action Network press release, November 10, 2004.

183 **Save Happy Valley Coalition occupation of Solid Energy headquarters:** "Anti-Coal Protestors Lock On to Solid Energy," Aotearoa Independent Media Centre, March 6, 2005.

184 **Mountain Justice Summer protest at National Coal Corporation:** "First Stockholders Meeting of National Coal Corporation Disrupted," Mountain Justice website, June 7, 2005.

184 **West Virginia citizens occupy Massey headquarters:** "Coalfield Citizens Arrested Delivering Demands to Massey Headquarters," Mountain Justice website, June 30, 2005; "Coalfield citizens arrested delivering demands to Massey headquarters," *Richmond Times Dispatch,* June 30, 2005.

184 **First Nations Mount Klappan mine blockade:** "Mine Road Blockade Rooted in Tahltan Dispute," *Toronto Globe & Mail,* September 7, 2005; "Fortune Minerals Blocked From Entering the Mount Klappan Coal Fields," Ontario Coalition Against Poverty website, July 28, 2005; "Struggles of the Tahltan Nation," *Canadian Dimension,* December 2005.

184 **Save Happy Valley Coalition coal train blockade:** "Protesters Stop Solid Energy Coal Trains," Save Happy Valley Coalition press release, August 13, 2005; "Save Happy Valley Members in Court," Save Happy Valley Coalition press release, February 9, 2006.

185 **Earth First! and Mountain Justice Summer blockade of Campbell County mountaintop removal site:** "Eleven Arrested Protesting ET Coal Mine," WBIR website, August 15, 2005; "Tennessee Coal Road Blocked to Protest Mountaintop Removal Mining," Mines & Communities website, August 15, 2005.

185 **Rising Tide boat blockade of Newcastle, Australia, port:** "Exporting Climate Disaster: People Take Back the Port," *It's Getting Hot In Here*, June 8, 2006.

185 **Earth First!/Rising Tide blockade of Clinch River Power Plant:** "Earth First! Blockades Power Plant," *Asheville Global Report*, July 26, 2007; "Resisting King Coal," Rising Tide website, July 11, 2006.

185 **Drax Power Station blockade attempt:** "In the Shadow of Drax, Not So Much a Fight as a Festival," *The Guardian*, September 1, 2006; "Green Protestors Mass to Close 'Drax the Destroyer'," Climate Ark website, August 31, 2006.

186 **Doodá Desert Rock blockade:** "Protesters Blockade Desert Rock Site," *Gallup Independent*, December 13, 2006; "Resisters Move," *Gallup Independent*, December 22, 2006.

186 **Rising Tide blockade of New South Wales Labor Party:** "Two Arrests in Coal Protest," Fox News, February 27, 2007; "ALP HQ Blockade: Decision on New-castle Coal Export Terminal Needed," Rising Tide Australia website, accessed January 2008.

186 **Sit-in at West Virginia governor Joe Manchin's office:** "11 Protesters Arrested at West Virginia Governor's Office," Mountain Justice Summer website, accessed January 2008.

186 **Blockade of Asheville Merrill Lynch:** "Climate Justice League Strikes Merrill Lynch," April 13, 2007.

187 **ASEN Blockade of New South Wales Department of Planning:** "Polar Bear Locks On at Department of Planning Against Anvil Hill Mine," Australian Student Environment Network website, June 8, 2007.

187 **Greenpeace blockade of New South Wales Department of Planning:** "Green-peace Dawn Blockade: Climate Protestors Call for 'No New Coal' as NSW Water Crisis Worsens," Greenpeace Australia press release, July 3, 2007.

187 **Southeast Convergence for Climate Action occupation of Asheville Bank of America:** "Protestors, Police Amass in Downtown Asheville," *Mountain Xpress*, August 13, 2007; "Southeast Convergence for Climate Action Shuts Down Bank of America," Blue Ridge Earth First! website, August 14, 2007.

187 **Occupation of Loy Yang Power Plant:** "Climate Protest Shuts Down Power Station," ABC News, September 3, 2007; "Disrupting Loy Yang," Real Action on Climate Change blog, September 3, 2007.

187 **ASEN Occupation of Newcastle coal port:** "11 Arrested at APEC Coal Protest," ASEN website, September 4, 2007.

188 **Greenpeace occupation at Boxburg plant construction site:** "German Coal Plant Construction Site Occupied," Greepeace International, October 3 and 4, 2007.

188 **Greenpeace occupation of Kingsnorth Power Plant:** "Protestors Raid Coal Power Plant," BBC News, October 8, 2007; "Greenpeace Shuts Down Coal Fired Power Station," Greenpeace UK website, October 8, 2007.

188 **Rainforest Action Network banner hang at Bank of America corporate headquarters:** "Charlotte Banner Tells Bank of America: Stop Funding Coal!" Rainforest Action Network UnderStory blog, October 23, 2007.

188 **Rising Tide boat blockade of Newcastle port:** "Protestors Block Coal Ships in Newcastle," *Sydney Morning Herald*, November 3, 2007; "Port Blockade a Success," Rising Tide Australia website, accessed January 2008.

189 **Rainforest Action Network activists and allies blockade a Citibank branch in Washington, DC:** "Coalfield Residents, Activists and Students Close Down D.C. Citi Branch," Rainforest Action Network press release, November 5, 2007.

189 **Rainforest Action Network Day of Action Against Coal Finance:** "Thousands Take to the Streets to Protest Citi and Bank of America's Coal Investments," Rainforest Action Network press release, November 16, 2007.

189 **Student blockade of Duke Energy headquarters:** "Students Chain Selves to Duke," *Raleigh News & Observer*, November 16, 2007; "Direct Action at Duke Energy Over Proposed Coal Expansion," *It's Getting Hot In Here*, November 15, 2007.

189 **Greenpeace occupation of Munmorah Power Station.** "Chain Reaction: 15 Protestors Arrested," *Sydney Morning Herald*, November 15, 2007.

190 **Rising Tide Kooragang Coal Terminal rail blockade:** "Coal Terminal Blockage Ended," *Sydney Morning Herald*, November 19, 2007; "Coal Train Blockaded," Rising Tide website, November 19, 2007.

190 **Blockade of Ffos-y-fran coal mine construction site:** "Activists Stop Welsh Coalmine Excavation," *The Guardian*, December 5, 2007.

190 **Mountain Justice Spring Break action at AMP-Ohio headquarters:** "Mountain Justice Takes On King Coal in Columbus," WattHead blog, March 28, 2008.

190 **Rising Tide and Earth First! occupation of Cliffside construction site:** "Eight Climate Protesters Arrested at U.S. Coal Plant," Reuters, April 1, 2008; "Eight Arrested as North Carolina Residents Shut Down Construction at Cliffside Coal Plant," Fossil Fools Day blog, April 1, 2008.

191 **Rainforest Action Network blockade of a Citibank office in New York City:** "Billionaires for Dirty Energy Blockade Citibank in New York, Two Arrested," Fossil Fools Day blog, April 1, 2008.

191 **Rising Tide and Rainforest Action Network blockade of Boston Bank of America branch:** "An April Fools'Protest," *Boston Globe*, April 1, 2008; "Activists Blockade Bank of America to Protest Funding of Coal, Boston," Fossil Fools Day blog, April 1, 2008.

191 **Occupation of Ffos-y-fran coal mine construction site:** "Making a Stand," *Merthyr Express*, April 3, 2008; "Protestors Shut Down Open-Cast Mine in Wales, Two Arrests," Fossil Fools Day blog, April 1, 2008.

191 **Eastside Climate Action blockade of E.ON headquarters, Nottingham, United Kingdom:** "Climate Protest in City Centre," *Nottingham Evening Post*, April 1, 2008; "Eastside Climate Action Blockade E.ON Workers As Part of Fossil Fools Day," *UK Indymedia*, April 1, 2008.

192 **Rising Tide occupation of Aberthaw Power Station:** "Aberthaw Power Station Successfully Blockaded This Morning," *UK Indymedia*, April 3, 2008; "Direct Action Double Whammy Against Welsh Carbon Dinosaurs," Luther ap Blissett blog, April 6, 2008.

192 **Blue Ridge Earth First! blockades Dominion Power's headquarters:** "3 Arrested in Protest Near Dominion Office," *Richmond Times Dispatch*, April 15, 2008; "We Shut Down a Major Corporation On an Hour of Sleep and So Can You!" Blue Ridge Earth First! website, April 15, 2008.

192 **Rising Tide blockade of coal terminal construction site in New South Wales:** "18 Arrested at Climate Change Protest," News.com.au, April 19, 2008.

192 **Activists halt coal train on its way to United Kingdom's largest power plant:** "Coal Train Ambushed Near Power Station in Climate Change Protest," *The Guardian*, June 14, 2008; "Police Arrest 29 Coal Train Protesters," Reuters UK, June 14, 2008.

193 **Protesters upstage Brisbane coal conference:** "Protesters picket Qld coal conference," Australian Broadcasting Corporation, June 16, 2008.

193 **Activists demonstrate outside Bank of America headquarters:** "Bank of America's Coal Investments Revisited," Rainforest Action Network Understory blog, June 26, 2008.

193 **Activists blockade Dominion headquarters:** "Thirteen Arrested in Protest at Dominion Today," *Richmond Times-Dispatch*, June 30, 2008.

193 **Greenpeace activists shut down a portion of Australia's most polluting power station:** "Activists Protest at Australian Power Plant," Reuters UK, July 3, 2008.

193 **Earth First! activists lock down at American Municipal Power headquarters, Columbus, Ohio:** "Women Climb Flagpole in Power Plant Protest," *NBC*

4 *Columbus,* July 7, 2008; "Police arrest protesters at Ohio power company," *WDTN 2 Dayton,* July 8, 2008.

194 **Mountain Justice activists protest approval of coal gasification plant, Boston, Massachusetts:** "Youth Protest State's Approval of Coal Gasification Plant," Mountain Justice website, July 10, 2008.

194 **Greenpeace activists occupy coal-fired power plant smokestack for thirty-three hours:** "Greenpeace Protesters Scale 140m Chimney," *Sydney Morning Herald,* July 11, 2008; "It's a Wrap: Watch a Video of the Occupation as It Unfolded," Greenpeace Australia website, accessed July 15, 2008.

194 **Blockades at Kooragang and Carrington coal terminals:** "Protest halts coal train for six hours," *Sidney Morning Herald,* July 14, 2008; "More coal protest arrests at Newcastle," *Business Spectator,* July 14, 2008; "Time for Action! People take action to halt coal exports," Camp for Climate Action, Australia, website, July 14, 2008.

195 **UK activists target coal-fired plant's PR agency:** "Activists Target Edelman in Climate Change Protest," *PR Week UK,* July 17, 2008; "Carbon Capture at E.ON's Kingsnorth Coal Plant," *Carbon Commentary,* January 14, 2008; "Oxford Climate Action Spin the Spinners!" *UK Indymedia,* July 16, 2008.

195 **Four arrested at Tennessee strip mine:** "Appalachian Residents Gather to March on Zeb Mountain," *The Small Ax,* July 21, 2008; "More Anti-Coal Direct Action at Appalachian Mine Site," Rainforest Action Network, July 21, 2008.

195 **Australian citizens blockade farm to stop coal exploration:** "People Power vs. Government Greed," Caroona Coal Action Group website, July 22, 2008; "Bring It On: Caroona Says 'No' to the Big Australian," *Northern Daily Leader,* July 22, 2008.

196 **Greenpeace paints anti-coal messages on twenty coal ships:** "Garrett Defends Coal Exports," *Sydney Morning Herald,* July 28, 2008.

196 **Activists glue themselves to coal giant's headquarters:** "Coal Protest Team Glued to Doors," BBC News, August 11, 2008.

196 **Southeast Convergence for Climate Action locks down at Bank of America, Richmond, Virginia:** "50 Protesters Urge Energy Regulation," *Richmond Times-Dispatch,* August 11, 2008; "SE Convergence Locks Down at Richmond Bank of America," *It's Getting Hot in Here,* August 11, 2008.

196 **Greenpeace Rainbow Warrior launches "Quit Coal" protest campaign in Israel:** "Israeli Police Arrest Greenpeace Rainbow Warrior Captain on 'Quit Coal' Protest," Greenpeace website, accessed October 6, 2008.

196 **Twenty protesters lock down at Dominion coal plant construction site in Wise County, Virginia:** "Peaceful Protesters Lock Their Bodies to Dominion Power Plant," Wise Up Dominion press release, September 15, 2008; "Dominion CEO Punk'd!" Rainforest Action Network Understory blog, September 15, 2008.

197 **Prime minister's office occupied:** "PM's Brisbane Office Targeted By Green Protesters," *Brisbane Times*, September 22, 2008; "Climate Protesters Occupy PM's Office," news.com.au, September 22, 2008

197 **Protesters shut down a Citibank branch in Cambridge, Massachusetts:** "Protestors Tell Citi and Bank of America: Not With Our Money, End Your Destructive Investments," Rainforest Action Network, September 27, 2008.

197 **Greenpeace "Quit Coal" tour visits Spain, boards coal ship:** "'Quit Coal' Action Against Coal Ship in Spain," *Scoop World Independent News*, October 6, 2008.

197 **Citizens rally at state capitol against new coal use, Little Rock:** "Arkansans Protest Against New Coal Use," KATV-6, October 18, 2008.

197 **Premier of Queensland's office occupied:** "Friends of Felton Occupy Anna Bligh's Office," Friends of Felton website, accessed November 12, 2008.

198 **Zombie March on top coal investors, Boston, Massachusetts:** "Zombie March on Coal's Top Investors, Copley Square, Boston," Ian MacLellan's Photo Blog, October 31, 2008.

198 **Rising Tide activists shut down Bayswater Power Station, New South Wales:** "25 Arrested at NSW Power Station Protest," *Sydney Morning Herald*, November 1, 2008; "29 Arrested After Six-Hour Climate Protest at Bayswater Power Station," Rising Tide Newcastle, November 1, 2008.

198 **Activists shut down Collie Power Station, Western Australia:** "Protesters Chained to Collie Power Conveyor," *West Australian*, November 5, 2008.

198 **Activists shut down Hazelwood power station:** "Hazelwood Tops List of Dirty Power Stations," World Wide Fund; "Protesters Target Hazelwood Power Station," Australian Broadcasting Corporation, November 6, 2008.

199 **Activists shut down Tarong Power Station, Queensland, Australia:** "Climate Activists Disrupt Australian Power Plant," Reuters, November 10, 2008; "Police Arrest Tarong Power Protesters," Australian Broadcasting Corporation, November 7, 2008.

199 **National Day of Action Against Coal Finance (November 14-15, 2008):** "Young Activists Fired Up in Fight Against Coal," *Post and Courier*, November 19, 2008.

199 **Greenpeace activists protest outside mine, Poznan, Poland:** "Polish Miners, Greens Clash on Eve of Climate Talks," *Planet Ark*, November 25, 2008.

199 **Activist shuts down Kingsnorth Power Station in the United Kingdom:** "Oldest Power Station 'Must Be Closed'," *Central Coast Express Advocate;* "No New Coal—The Calling Card of the 'Green Banksy' Who Breached Fortress Kingsnorth," *The Guardian*, December 11, 2008.

199 **Santa Protest at TVA in Knoxville:** "Santa Protests TVA," dirtycoaltva.blogspot.com/2008/12/santa-protest-tva.html, December 6, 2008.

200 **Santa Detained at TVA in Chatanooga:** youtube.com/watch?v=YECmuMFNcG8.

200 **Sludge Safety Lobby Day:** Jeff Biggers, "Takes a Village to Stop Razing Appalachia: Power Past Coal Fights Back," March 12, 2009, Power Past Coal.

201 **Coal River Mountain activists arrested, Pettus, West Virginia:** Jeff Biggers, "Takes a Village to Stop Razing Appalachia: Power Past Coal Fights Back," Power Past Coal website, March 12, 2009; "Coal River Mountain Can't Wait," *Grist*, February 3, 2009; "Fourteen Arrested Defending Coal River Mountain," Power Past Coal website, February 3, 2009.

201 **Rising Tide Boston crashes Arch Coal CEO lecture, Cambridge, Massachusetts:** "Rising Tide Boston Crashes Talk by Arch Coal CEO," Power Past Coal website, February 5, 2009.

201 **Billionaires for Coal visit Dominion headquarters in Richmond, Virginia:** "Billionaires for Coal Visit Dominion Resources HQ in Richmond February 7," *RootsWire*, February 7, 2009.

201 **Ed Schultz action:** "Grassroots Rockin' on Ed Schultz!" GrassrootsGrow blog, February 9, 2009.

202 **Santee Cooper protest:** Jeff Biggers, "Takes a Village to Stop Razing Appalachia: Power Past Coal Fights Back," March 12, 2009, Power Past Coal.

202 **Activists close accounts with Bank of America, San Francisco, California:** "Rising Tide Bay Area: 'Bank of America, Where's Your Heart?'" Rainforest Action Network Understory blog, February 15, 2009.

202 **Two arrested for halting blasting at mountaintop removal site, Raleigh County, West Virginia:** "Blasting at Clays Branch" Climate Ground Zero press release, February 16, 2009.

202 **Frankfurt protest:** "Hundreds Call for End to Mining Damage," *Courier-Journal*, February 18, 2009.

202 **March in Corpus Christi, Texas:** "Hundreds Call for End to Mining Damage," *Courier-Journal*, February 18, 2009.

203 **Activists rally against coal in Massachusetts"** "Action en Mass!" Power Past Coal website, March 8, 2009.

203 **Thousands gather to protest coal and global warming, Washington, DC:** "Thousands Rally for Legislation on Climate Change," Associated Press, March 2, 2009; "Pelosi/Reid Call to Switch Capitol Power Plant off of Coal!" *It's Getting Hot in Here*, February 26, 2009; "Thousands Storm Capitol Hill in Largest Protest Against Global Warming," *AlterNet*, March 3, 2009.

203 **United Mountain Defense volunteer arrested by TVA:** "Volunteer Arrested at TVA Ash Disaster Site," Student Environmental Action Coalition, March 6, 2009, seac.org/node/174.

203 **Activists protest mountaintop removal, Pettus, West Virginia:** "Operation Appalachian Spring: Sit-in Coal Campaign Blooms," Power Past Coal press release, March 5, 2009.

204 **Middlebury "Freeze on Coal":** Jeff Biggers, "Takes a Village to Stop Razing Appalachia: Power Past Coal Fights Back," March 12, 2009, Power Past Coal website.

204 **Council Building blockade in Brussels, Belgium:** "Over 300 Greenpeace Activists Arrested After Finance Ministers Blockade," Greenpeace press release, March 10, 2009.

204 **Protesters march against coal in Palm Springs, California:** "March For Clean Energy Seeks to Curb Coal Use," *Desert Sun,* March 15, 2009.

204 **Fourteen arrested at TVA headquarters in Knoxville, Tennessee:** "March 14, 2009: Fourteen Arrested at TVA Headquarters During March in March," Mountain Justice website, March 15, 2009; youtube.com/watch?v=8hZhjd2dNBg.

205 **Anti-coal protesters gather outside statehouse in Topeka, Kansas:** "Anti-coal Groups Converge on KS Statehouse," *Kansas City Star,* March 19, 2009.

205 **"Bluegrass at the Bank" hits Bank of America branch in Sarasota, Florida.** "Bluegrass at the Bank Strikes Again!" Mountain Justice website, March 20, 2009.

205 **Protesters blockade coal terminal in Newcastle, Australia:** "Protesters Close Newcastle Coal Terminal," *Steel Guru,* March 23, 2009.

205 **Rising Tide disrupts coal-to-liquids conference in Washington, D.C.:** "DC Rising Tide Disrupts Coal-to-Liquids Conference," *It's Getting Hot in Here,* March 26, 2009.

206 **Students rally outside capitol in Austin, Texas:** "ReEnergize Texas!" Power Past Coal website, March 31, 2009.

206 **Inanimate activist with Mannequins for Climate Justice shuts down Bank of America branch in Boston, Massachusetts:** "Mannequins for Climate Justice Shut Down Bank of America," Power Past Coal website, March 31, 2009.

206 **Greenpeace activists hold a "coal circus" on Boston Common:** "Clean Coal? April Fools Says Greenpeace," *The Boston Globe,* April 1, 2009.

206 **Over a hundred arrested for allegedly planning direct action against coal plant in Nottingham, United Kingdom:** Juliette Jowit and Matthew Taylor, "Mass Arrests Over Power Station Protest Raise Civil Liberties Concerns," *The Guardian,* April 14, 2009.

206 **Activists arrested at Massey Energy mine in West Virginia:** "Activists Hang "EPA Stop MTR" Banner on Massey Mine, Arrested," Climate Ground Zero press release, April 16, 2009.

207 **Hundreds protest in Charlotte, North Carolina, against Duke's proposed Cliffside plant:** "Hundreds March and 44 Arrested to Stop Cliffside Power Plant," Power Past Coal press release, April 21, 2009.

207 **Activists begin fast to urge immediate action on global warming.** Fasting for our Future website, accessed May 4, 2009.

207 **Greenpeace activists hang banner at international climate meeting in Washington, D.C.:** "It's a Beautiful Day for a Banner Hang!," Greenpeace USA website, April 27, 2009.

207 **Activists protest Cliffside plant at Duke Energy shareholder meeting:** "Coal Debate Highlights Duke Meeting," *Triangle Business Journal,* May 8, 2009.

207 **Police remove eleven activists from mountaintop removal protests in West Virginia** "Removal Coal Mining; More Protestors Expected This Afternoon," Press Release, May 23, 2009; "Group Raising Money for Bail for Coal Protesters," Associated Press, May 26, 2009.

208 **Brushy Fork arrests:** "Non-violent Civil Disobedience in Coal River Valley, WV: Seventeen Arrested in Three Separate Actions," Mountain Justice website, May 23, 2009.

208 **Activists board coal ship in Kent, England:** "Protesters Leave Coal Cargo Ship," BBC News, June 22, 2009. "Thousands demonstrate against coal plant in Mainz," The Local, May 24, 2009.

209 **Mainz, Germany protest:** "Thousands demonstrate against coal plant in Mainz," *The Local,* May 24, 2009.

209 **Activists scale 20-story dragline at mountaintop removal site in Twilight, West Virginia:** Jeff Biggers, "Daring Dragline Protest Launches 7 Days That Will Shake Mountaintop Removal Operations," *Common Dreams,* June 18, 2009.

209 **Activist board ship in Kent, England:** "Protesters leave coal cargo ship," BBC News, June 22, 2009.

209 **Dozens arrested protesting at Massey Energy site in Coal River Valley, West Virginia:** Jeff Biggers, "Nonviolent Goldman Prize Winner Attacked by Massey Supporter: 94-Year-Old Hechler, Hannah, Hansen Arrested at Coal River," *Huffington Post,* June 23, 2009.

209 **More than 700 turn out against carbon sequestration:** "700 protest against carbon dioxide plan," *Dayton Daily News,* June 30, 2009.

210 **Banner drop at EPA, Boston:** "Boston Rising Tide Activists Drape Banner On EPA Building, Call on EPA to Stop Mountaintop Removal," Rainforest Action Network Understory blog, June 29, 2009.

210 **Mount Rushmore:** "Greenpeace Activists Arrested After Draping Banner on Mount Rushmore," *Washington Post,* July 8, 2009.

210 **Greenpeace activists spray-paint coal ship and power station in Italy:** Jani Myer, "Greenpeace Red-Flags Ship Carrying SA Coal," *Independent Online,* July

11, 2009; "BHP Coal Berth Blocked by Greenpeace Ship as Protest Continues," *Bloomberg*, August 6, 2009.

210 **Boulder rally:** "Public packs Valmont power plant hearing," Clean Energy Action, July 14, 2009.

210 **Lansing rally:** "Crowd rallies at Capitol for renewable energy," *Lansing State Journal*, July 29, 2009.

211 **Hay Pt. terminal blockade:** "BHP Coal Berth Blocked by Greenpeace Ship as Protest Continues," Bloomberg, August 6, 2009.

211 **Hamilton, UK protest:** "Coal protest outside council HQ," BBC News, August 10, 2009.

211 **Charleston, WV lockdown:** "4 Lockdown at WV Dept. of Environmental (No) Protection," *It's Getting Hot In Here*, August 11, 2009.

211 **National Coal "Going Away Party":** "Hug and Love National Coal Number 2 protest 8 0001," youtube.com/watch?v=9JFFrJ9NCoU; "Love and Hug National Coal Protest 3 Stockholders 8 20 09," youtube.com/watch?v=Kcu5dl8blPQ.

212 **Activists occupy trees in Coal River Valley:** "Treesit stopping blasts above Pettry Bottom, Coal River Valley," Climate Ground Zero, August 25, 2009; "Tree-sitting protest of mountaintop removal ends in W.Va. after 6 days; activists arrested," Taragana blog, August 31, 2009.

APPENDIX B

Coal Plants Canceled, Abandoned, or Put on Hold

For further details on particular plants, see "Stopping the Coal Rush," Sierra Club new coal plant proposal tracking list, sierraclub.org/environmentallaw/coal/plantlist.asp, or CoalSwarm's wiki pages organized by year: "Coal plants cancelled in 2007," sourcewatch. org/index.php?title=Coal_plants_cancelled_in_2007; "Coal plants cancelled in 2008," sourcewatch.org/index.php?title=Coal_plants_cancelled_in_2008; "Coal plants cancelled in 2008," sourcewatch.org/index.php?title=Coal_plants_cancelled_in_2009;

213 **Hunter Unit 4:** "Stopping the Coal Rush," Sierra Club new coal plant proposal tracking list.

213 **Big Brown 3, Morgan Creek 7, Tradinghouse 3 and 4, Sandow 5, Monticello 4, Martin Lake 4, Lake Creek 3:** Andrew Ross Sorkin, "A buyout deal that has many shades of green," *New York Times*, February 26, 2007.

213 **Cliffside second unit:** "Stopping the Coal Rush," Sierra Club new coal plant proposal tracking list.

213 **Corn Belt plant:** "Stopping the Coal Rush," Sierra Club new coal plant proposal tracking list.

213 **Indian River Power Plant:** "Indian River," CoalSwarm wiki,

214 **Escanaba plant:** "WPPI Stows Earlier Coal Plant Proposal," *Platts Coal Outlook,* May 14, 2007.

214 **PacifiCorp plants:** See Chapter Nine, "The Education of Warren Buffett."

214 **Nueces IGCC plant:** "Stopping the Coal Rush," Sierra Club new coal plant proposal tracking list.

214 **Taylor Energy Center:** "Taylor Energy Center," CoalSwarm wiki.

214 **Glades Power Plant:** "Glades," CoalSwarm wiki.

214 **Sallisaw Electric Generating Plant:** "Stopping the Coal Rush," Sierra Club new coal plant proposal tracking list.

214 **LS Power Sussex proposal.** "Stopping the Coal Rush," Sierra Club new coal plant proposal tracking list.

214 **Thoroughbred Generating Station:** "Court Says No to Peabody Coal," *Media Island International,* August 9, 2007.

214 **Seminole 3 Generating Station:** "Seminole 3," CoalSwarm wiki.

214 **Nelson Creek Project:** "Nelson Creek Project," CoalSwarm wiki.

215 **South Heart Power Project:** "South Heart on Life Support," *Dakota Counsel,* August 2007.

215 **Mesaba Energy Project:** "State Agency Blocks Coal Plant," Minnesota Center for Environmental Advocacy, August 3, 2007.

215 **Holcomb Unit 3:** Tim Carpenter, "Holcomb Plant at Center of Emissions Conflict," *Topeka Capital-Journal,* September 23, 2007.

215 **Russell Station II:** Daniel Wallace, "Russell Station Plans Change," *Rochester Democrat and Chronicle,* September 29, 2007.

215 **Gascoyne 175 Project:** "MDU Shelving Gascoyne Power Plant," *Bismarck Tribune,* May 29, 2006.

215 **Roundup Power Project:** Clair Johnson, "Roundup Power Permit Invalid," *Billings Gazette,* July 17, 2007.

215 **Red Rock Generating Station:** "OCC Denies Application for Red Rock Plant," AEP website, September 10, 2007.

215 **Avista plant:** "Avista Issues Resource Plan," Avista press release, September 4, 2007.

215 **Bowie IGCC Power Station:** "Bowie IGCC Power Station," CoalSwarm wiki.

215 **Holcomb Units 1 and 2:** Steven Mufson, "Power Plant Rejected Over Carbon Dioxide for First Time," *Washington Post,* October 10, 2007.

216 **Marion Gasification Plant:** Alex Klein, "TECO, Nuon Cancellations Underscore IGCC's Woes," Emerging Energy Research, October 5, 2007, Exhibit 2.

216 **Huntley Generating Station:** Alex Klein, "TECO, Nuon Cancellations Underscore IGCC's Woes," Emerging Energy Research, October 5, 2007, Exhibit 2.

216 **Buffalo Energy Project:** Alex Klein, "TECO, Nuon Cancellations Underscore IGCC's Woes," Emerging Energy Research, October 5, 2007, Exhibit 2.

216 **Xcel IGCC plant:** "Unnamed Xcel Energy Plant," CoalSwarm wiki.

216 **Polk Power Station:** "Polk Power Station Unit 6," CoalSwarm wiki.

216 **West Deptford Project:** "West Deptford Project," CoalSwarm wiki.

216 **Stanton Energy Center:** "Stanton Energy Center," CoalSwarm wiki.

216 **Pacific Mountain Energy Center:** "Pacific Mountain Energy Center," CoalSwarm wiki.

216 **Twin River Energy Center:** "Twin River Energy Center," CoalSwarm wiki.

217 **Matanuska Power Plant:** "Matanuska Power Plant," CoalSwarm wiki.

217 **Idaho Power project:** John Miller, "Idaho Power gives up on coal-fired plant," *Idaho Statesman,* November 7, 2007.

217 **Elmwood Energy Center:** "Elmwood Energy Center," CoalSwarm wiki.

217 **Rentech Energy Midwest:** "Rentech Energy Midwest Corporation," CoalSwarm wiki.

217 **Alcoa project:** Alan Brody, "Power Plant Plug Pulled," *Southern Maryland News Online,* December 14, 2007.

217 **Jim Bridger Station expansion:** "PacifiCorp Cancels Wyoming Coal Projects," *Wyoming Tribune-Eagle,* December 11, 2007.

217 **Southern Illinois Clean Energy Center:** "Stopping the Coal Rush," Sierra Club status list.

217 **Soda Springs project:** "Stopping the Coal Rush," Sierra Club status list.

217 **Jim Bridger IGCC demonstration project:** "PacifiCorp Cancels Wyoming Coal Projects," *Wyoming Tribune-Eagle,* December 11, 2007.

218 **Intermountain Power Project expansion:** Jasen Lee, "PacifiCorp to fuel plant with wind, gas—not coal," *Deseret Morning News,* December 8, 2007; "Intermountain Power Project Unit 3," CoalSwarm wiki.

218 **Kansas City Board of Public Utilities project:** "Stopping the Coal Rush," Sierra Club new coal plant proposal tracking list.

218 **Bethel Power Plant:** "Bethel Power Plant," CoalSwarm wiki.

218 **Rosemount Project:** "Stopping the Coal Rush," Sierra Club new coal plant proposal tracking list.

218 **Ray D. Nixon Power Plant:** "Stopping the Coal Rush," Sierra Club new coal plant proposal tracking list.

218 **Fayette County Economic Development Project:** "Fayette County Economic Development Project," CoalSwarm wiki.

218 **Baldwin Energy Complex:** "Baldwin Energy Complex," CoalSwarm wiki.

218 **Illinois Energy Group project:** "Illinois Energy Group," CoalSwarm wiki.

219 **Elkhart Proposal (Turris):** "Elkhart Proposal (Turris Coal)," CoalSwarm wiki.

219 **AES Colorado Power Project:** "AES Colorado Power Project," CoalSwarm wiki.

219 **High Plains Energy Station:** "High Plains Energy Station," CoalSwarm wiki.

219 **Buick Coal and Power Project:** "Buick Coal and Power Project," CoalSwarm wiki.

219 **FutureGen:** "FutureGen," CoalSwarm wiki.

219 **Big Cajun II Unit 4:** "NRG Energy, Inc. Q4 2007 Earnings Call Transcript," Seeking Alpha website, Feb. 28, 2008.

219 **Kenai Blue Sky Project:** "Kenai Blue Sky Project," CoalSwarm wiki.

219 **Norborne Baseload Plant:** "Norborne Baseload Plant," CoalSwarm wiki.

220 **Mountaineer IGCC:** "Mountaineer," CoalSwarm wiki.

220 **SIU Power Plant:** "Stopping the Coal Rush," Sierra Club new coal plant proposal tracking list.

220 **Gascoyne 500 Project:** "Coal Company Suspends Effort to Build N.D. Power Plant," *Bismarck Tribune,* May 21, 2008.

220 **Milton Young 3:** "Minnkota, FPL Announce Wind Farm Plans," Minnkota press release, March 29, 2007; Lauren Donovan, "Economic Pinwheels Spinning in Oliver," *Bismarck Tribune,* June 25, 2008.

220 **Gilberton Coal-to-Clean-Fuels and Power Project:** "Gilberton Coal-to-Clean-Fuels and Power Project," CoalSwarm wiki.

220 **Sithe Shade Township Project:** "Sithe Shade Township Project," CoalSwarm wiki.

221 **Lower Columbia Clean Energy Center:** "Lower Columbia Clean Energy Center," CoalSwarm wiki.

221 **Twin Oaks Power Unit 3:** "PNM Resources Drops Texas Coal-Plant Expansion," Reuters, August 12, 2008.

221 **Western Greenbrier Co-Production Demonstration Project:** "Western Greenbrier Co-Production Demonstration Project," CoalSwarm wiki.

221 **Buffalo Energy Project:** "Buffalo Energy Project," CoalSwarm wiki..

221 **Benwood Plant:** "Benwood Project," CoalSwarm wiki.

221 **Nelson Dewey Generating Facility expansion:** "PSC Rejects Alliant Energy's Proposed Coal Plant," *Journal Sentinel,* November 11, 2008.

222 **Indiana SNG:** "Indiana SNG," CoalSwarm wiki.

222 **Kentucky Mountain Power:** "Kentucky Mountain Power," CoalSwarm wiki.

222 **Lima Energy Plant.** "Stopping the Coal Rush," Sierra Club new coal plant proposal tracking list.

222 **Thoroughbred:** "Thoroughbred Generating Station," CoalSwarm wiki.

222 **Elk Run Energy Station:** "Elk Run Energy Station," CoalSwarm wiki.

222 **Highwood Generating Station:** "Highwood Generating Station," CoalSwarm wiki.

222 **Malmstrom Air Force Base Coal-to-Liquids:** "Malmstrom Air Force Base Coal-to-Liquids," CoalSwarm wiki.

223 **AES Shady Point:** "AES Shady Point II," CoalSwarm wiki.

223 **Ely Energy Center:** "Ely Energy Center, Phase I," CoalSwarm wiki.

223 **Little Gypsy Repowering project:** "Little Gypsy Repowering," CoalSwarm wiki.

223 **White Pine Energy Station:** "White Pine Energy Station," CoalSwarm wiki.

223 **Sutherland Generating Station Unit 4:** "Sutherland Generating Station Unit 4," CoalSwarm wiki.

223 **Unnamed Tri-State plant:** "Tri-State to Review Coal-Fired Power Plant Prospects," *Power Engineering*, April 13, 2009.

223 **NextGen Energy Facility:** Letter from Lyle Witham to Lynn Gustafson, South Dakota Department of Environment and Natural Resources website, May 18, 2009.

224 **Midland Power Plant:** Tina Lam, "Plan for Coal Plant in Midland Cancelled," FreeP.com.

224 **Northern Michigan University Ripley Addition:** "Stopping the Coal Rush," Sierra Club new coal plant proposal tracking list.

224 **Intermountain Power Project Unit 3:** Jasen Lee "PacifiCorp to Fuel Plant with Wind, Gas—Not Coal," *Deseret Morning News*, December 8, 2007.

■

Index

■

ABOUT COALSWARM

Citizens are mobilizing. In the United States and around the world, over 375 groups are now working on coal issues. Most of these are locally based organizations whose effectiveness is often underestimated. The mission of CoalSwarm is to assist this movement by building shared resources. The CoalSwarm wiki (http://CoalSwarm.org), created in collaboration with the Center for Media and Democracy, provides a constantly expanding information clearinghouse on coal, including over 2,000 articles. Anyone can contribute information, and scores of people have participated in developing the website. The only requirement is that all facts must be linked to published sources. The following are some of the resources provided by CoalSwarm:

Coal plants and mines
- Proposed coal plants (over 375 projects in 26 countries)
- Existing coal plants (over 600 facilities)
- Coal plant cancellations
- Campus coal plants
- Coal plant conversion projects
- Mines

Citizen activism
- Citizen groups involved in coal issues:
- Nonviolent direct action and other protests
- Coal issues calendar
- Coal plant litigation
- Coal activist videos
- Expert testimony

Coal-related companies and agencies
- Power companies and agencies
- Rural electric cooperatives
- Mining companies
- Synfuels companies
- Railroads

Coal and power industry data
- Coal reserves
- Coal exports
- Plant capacity and output statistics
- Emissions data
- Coal and jobs

Politics and coal
- Coal money in politics
- Lobby groups and trade associations
- Obama administration statements and policies

Coal-related environmental issues
- Coal waste
- Mountaintop removal
- Air emissions

Alternatives to coal
- Solar, wind, geothermal, efficiency; comparative economics; legislation and policy options

ABOUT THE COVER

The cover photograph, which depicts an unidentified protester approaching a twenty-story dragline in an attempt to hang a banner reading "Stop Mountaintop Removal: Clean Energy Now!" is excerpted from video footage recorded by filmmaker Kurt Mann before dawn on June 19, 2009, at Massey Energy's Twilight mine in Boone County, West Virginia.

Fourteen protesters affiliated with Rainforest Action Network and Climate Ground Zero entered the mine, one of the largest mountaintop removal coal mines in North America. Mann and the other protesters were arrested for trespassing after occupying the dragline. Mann was released on bail later in the day but his camera gear and media of the event were detained as "evidence." Nevertheless, Mann managed to have some extraordinary footage taken off site when it was clear the police would force filming to stop.

Mann's documentary *Planet Ground Zero,* produced by American Green, is scheduled for release in 2010.

The Twilight mine produced over five million tons of coal in 2007. Coal from the mine is transported via underground conveyer to the Elk Run Resource Group for processing and shipment on the CSX railway.